EXPERIMENTAL CHEMISTRY

Laboratory Manual for Chemistry 14

Staff of the Chemistry Department
The Pennsylvania State University

Edited by

Charles G. Haas

The Pennsylvania State University
University Park, Pennsylvania

KENDALL/HUNT PUBLISHING COMPANY

2460 Kerper Boulevard,
Dubuque, Iowa 52001

Printed in the United States of America
10   9   8   7   6

Contents

EXPERIMENTAL SECTION

Preface

The material in this manual has evolved during several years' use by the Chemistry Department at all campuses of the Pennsylvania State University. Our goal has been to develop a challenging and suitable set of quantitative experiments for students in a first-year university chemistry course. The course stresses the central nature of quantitative relationships in the physical sciences and the necessity for careful experimentation, ideas which are important for all students of chemistry. For majors in chemistry and chemically allied fields, the course also provides an introduction to some of the analytical techniques formerly encountered in a separate course in quantitative analysis. Since we believe that the ability to maintain a reliable record of work done is important in a wide variety of science and non-science disciplines, we have placed considerable emphasis on the laboratory notebook and the analysis of data.

A departmental course like Chemistry 14 is the result of the cooperative interaction of many individuals. When the laboratory program was modernized over a decade ago, the manual for the new content was produced largely by T.V. Long. As the program was implemented, suggestions for changes and improvements were offered by numerous members of the Chemistry Department at the Commonwealth Campuses and at University Park. In 1976 a detailed review of Chemistry 14-15 and a revision of the laboratory manuals was completed by a committee chaired by J.P. Lowe and consisting of G.L. Geoffroy, T.H. Risby, and the current editor.

In preparing for this first commercial printing of the manual many additional changes have been made. Much of the introductory material has been completely rewritten and considerable portions of the experimental write-ups have been revised in light of recent student feedback. The sections on Errors and Their Analysis is based largely on notes prepared earlier by Professor Sami Talhouk. Senior chemistry student William Levinson presented the organization and most of the examples for the summary of error terminology. Thanks are due to these and to many others who indicated places where clarity could be improved.

The history of the continuous evolution of Chemistry 14 which has preceded this version of our manual will undoubtably extend into the future. We ask all the users to continue offering suggestions, criticisms, and corrections to improve the utility to future students and instructors.

Introductory Section

TO THE STUDENT

You are beginning a first course in quantitative laboratory chemistry. For many this course will be a departure from former laboratory experiences in that you will be expected to obtain precise numerical results for every experiment which is performed. If an experiment is botched up somewhere along the way or if results are obviously poor, the procedure should be analyzed for sources of difficulty and the experiment should be repeated to obtain better data. Generally there will be sufficient time for repetition of poor determinations, but you may check with an instructor if there is doubt.

Each experiment should be read before coming to laboratory. To promote this practice, weekly (10-20 minutes) quizzes on the current experiment will often be administered. The content of these will generally be related to the following aspects of the experiment:

a. the purpose of the determination;
b. the procedure;
c. the data to be obtained;
d. the possible sources of error and danger.

Questions given at the end of each experimental write-up will give you some idea of the kinds of questions which might be asked in quizzes.

The experimental write-ups are provided with references to more detailed discussions of various aspects of the experiment. Textbooks are referred to by author. The full reference for each such text is given below.

T.L. Brown and H.E. LeMay, "Chemistry—The Central Science," Prentice Hall, Inc. (1977).
E.J. Masterson and W.L. Slowinski, "Chemical Principles," W.B. Saunders, Co., 4th edition (1977).
C.E. Mortimer, "Chemistry, A Conceptual Approach," Van Nostrand Reinhold Co., 3rd edition (1975).
M.J. Sienko and R.A. Plane, "Chemistry—Principles and Properties," McGraw-Hill Book Co., 2nd edition (1974).

The experiments in Chemistry 14 are designed to help you master some specific chemical concepts and procedures as well as to introduce you to certain specific techniques. Mastery of this body of chemical laboratory work is an important goal. There is, however, another primary goal of a more general sort—to develop a sound approach toward any kind of scientific experimentation. There are many aspects of this ability: careful experimentation, proper recording of data and observations, clear (as well as correct) presentation of calculations and results, adequate analysis of possible errors, etc. You will find that the comparable amounts of emphasis will be placed on good experimentation and on a good laboratory notebook record of the experiment.

The introductory sections of this manual provide some background material to help you in this course. You will find presentations on safety rules, keeping a laboratory notebook, experimental errors and their analysis, graphing, and the use of certain laboratory equipment. It is essential for you to read this introduction carefully before you begin laboratory work, and to refer back to it from time to time during Chemistry 14.

1

Safety and Laboratory Rules

Study, learn, and observe the rules given below.

1. **Safety goggles of approved design must be worn at all times in the laboratory,** even over personal eyeglasses if this be the case. Failure to do so is a violation of Pennsylvania State Law. *This rule will be enforced!* Anti-fogging agent will be available in the laboratory.

2. If you use *contact lenses,* be sure to remove and clean them as soon as you leave the laboratory. Small amounts of irritants, normally harmless, may concentrate between the lens and the cornea and cause serious problems if not removed promptly. [Wearers of contact lenses should be alert to this danger anytime they encounter air-borne irritants whether in a laboratory or elsewhere.]

3. During the first laboratory period familiarize yourself with the location of the safety features of the laboratory in which you are working. These include the *safety shower, fire extinguisher, fire blanket, eye wash bottle, laboratory first aid kits,* and the *first aid room.* The safety shower should be used if your clothing catches fire or if a corrosive chemical is spilled on you in quantities that cannot be easily flushed away at laboratory faucets.

4. You will be using a number of organic liquids in this course which may be flammable and/or poisonous. These include ethyl alcohol, acetone, and samples provided as unknowns. Ethyl alcohol and acetone and their vapors are very easily ignited by flame; thus extreme caution must be exercised to avoid having a flame in the immediate vicinity of containers of these liquids. In addition the vapors of some of the unknowns are slightly toxic. **For the above reasons unknown liquids are not to be dumped into lab sinks.** Capped containers for depositing the waste solutions from the experiments employing unknown organic liquids are provided in the lab. It will be acceptable for the acetone, used in pipet rinsing, etc., in several experiments, and *ethyl alcohol* used in topic 3, to be deposited in lab sinks, *accompanied by a thorough flushing with water.*

5. Eating, drinking, and smoking are **forbidden** in the laboratory. Be sure to wash your hands after leaving the laboratory before you handle food.

6. *Never* return unused reagents to the stock bottle, since this may contaminate a common supply. If you take too much of a liquid or soluble solid, dispose of it in the sink. Likewise *never* insert a pipet or a spatula into a supply bottle.

7. Deposit solid wastes like filter paper, used matches, etc., into the crocks or cans provided in the lab. Do not place solid wastes in the sinks, since they will plug the drains.

8. Always *lubricate* glass tubing and thermometers with water or glycerine (available on lab shelf) before inserting them into a stopper. Always wrap toweling around them while inserting. This will reduce the chance of breaking the glass and cutting your hand.

9. Use your *desk hood* whenever poisonous or irritating fumes are evolved in an experiment. Generally, best ventilation will be achieved in desk hoods if the mouth of the vessel is placed close to the top of the center of the hood (1 or 2 inches away).

10. The Bunsen burner can become a major fire hazard IN THE HANDS OF A CARELESS STUDENT. The burner should be lit and kept burning only for the period of time in which it is actually utilized. Carefully position it on the desk away from the reagent shelf. Before lighting your burner, make sure that flammable reagents on neighboring desks are well separated from your burner. Be careful not to put your arm over a lit burner while reaching for something. Long hair must be tied back to minimize the possibility of contact with burners or other equipment.

11. When heating or carrying out a reaction in a test tube, never point this test tube at your neighbor or yourself.

12. Never taste a chemical. When you are instructed to smell a chemical, point the container away from your face and use your hand gently to fan a little of the vapor to your nose.

13. Never pour water into concentrated acid. Always pour the *acid* slowly *into* the *water* while stirring constantly.

14. If you receive a chemical burn by acid, alkali, or other liquid, immediately flood the burned area with water. Then have another student summon the laboratory instructor. If you get a splash of any material in your eye, **immediately** flush it with water at the *nearest* sink. Hold the lids open with one hand while you splash water in with the other. Also, send another student for the instructor at once.

15. Always use a rubber bulb with your pipet. *Never pipet by mouth.*

16. Makeshift equipment and set-ups are the first steps to an accident. Always assemble your apparatus as outlined in your instructions.

17. Outdoor garments, books, etc., should not be placed on the lab bench where they interfere with experiments and are liable to damage.

18. Shoes must be worn at all times in the laboratory.

19. Under no conditions are unauthorized experiments to be performed in the laboratory.

20. *Report any accident, however minor, to the instructor at once.*

Notebooks and Reports

Keeping a Laboratory Notebook

The two principal activities of a scientist in the laboratory are making careful observations and accurately recording those observations. Hence, an important goal of any chemistry course is training students to keep accurate, complete, and permanent records of their laboratory work. You are expected to keep such a laboratory notebook, and it will be examined periodically by your instructor. Failure to maintain a notebook at the proper level of completeness and accuracy will adversely affect your grade.

A suitable notebook is made by Wm. H. Freeman Co., San Francisco, and is available at the bookstores. It contains tear-out pages and is designed to allow you to make a carbon copy of each page. (The original copy of each page will be collected by your instructor at the end of each laboratory period.)

The following guidelines should govern your notebook record.

1. Number each notebook page consecutively (throughout the notebook) in the upper right corner. (It is generally good practice to skip a couple of pages at the start of a notebook to allow for an index to be made.)
2. Identify each page as you start it by your name, experiment number, and date. If your notebook has blanks for these, enter them in the blanks. Otherwise they should be written in the upper left corner.
3. If a given page is not completed in one period and is used for the continuation of the experiment, the second date should appear at the start of the new work.
4. Pencil records are *not* acceptable. Use ball point pen, with carbon paper for the second copy.
5. All data and observations should be recorded **directly in the notebook** as the observation is being made. It is completely unacceptable to record raw data (buret readings, weights, etc.) on *any* other paper.
6. For experiments done by pairs of students, it may be more convenient to have only one partner record the data as the experiment is performed. In such cases the other partner should then copy the information into his notebook as soon as possible after the data are collected.
7. Each entry should be labelled so that it is clear what the number refers to. Correct units should be included with the number.
8. Headings describing the laboratory procedure associated with the ensuing data are helpful (e.g., "Standardization of $KMnO_4$ Solution," "Analysis of Unknown," etc.).
9. *Make no erasures* (almost impossible anyway, if you are using ink) or obliterations of data you have entered. It is important that your notebook be a complete record, which means that it should contain your mistakes along with your successes. Any erroneous entry should be neatly crossed out with a single horizontal line; the correction should be entered beside the original incorrect entry. But this should be done so as not to make the original numbers illegible. Sometimes one later decides he needs those numbers after all. If there is a sizable section which is invalid, say, because of an error, a diagonal line through the section is appropriate, with horizontal lines at the top and bottom of the discarded section.
10. An expository discourse of the procedure is *not* to be included. Comments, however, about unusual occurrences should be included where appropriate (e.g., "Discarded wrong solution by mistake—start over!").

11. In addition to numerical readings and calculations, be sure to include other important data which should be recorded but are often overlooked. Examples are: code numbers for unknown samples, room temperature or pressure for certain experiments, concentrations of side-shelf reagents, observations of colors or color changes, precipitate formation, etc.

Your instructor will give further suggestions and instructions in the keeping of a laboratory record. You should avoid producing a notebook wherein unlabelled data and calculations are scattered all over the page in a manner which only you can understand today and nobody can understand tomorrow. *Strive always to make your notebook record as legible as your handwriting permits.*

The Notebook as Laboratory Report

Your laboratory report is comprised of the notebook sheets you have turned in during and after the experimental work. Naturally, the earliest sheets are likely to contain mostly the data you have gathered and certain related calculations. Since these are the original laboratory records, they should **not be recopied** or otherwise changed. The later sheets are likely to contain tabulations, graphs, conclusions, discussions of errors.

The policy of using the original laboratory record sheets as a part of the experimental report means that it will pay to do some planning before you start an experiment. You will want to survey the write-up, determine the sort of data to be gathered, and to give some advance consideration to the way you organize your notebook for the experiment. Generally it will be convenient to organize the write-up into the following sections.

a. **Raw and derived data.** This would include the observed data, such as meniscus heights in burets, as well as data derived from these, such as volumes of liquids titrated. Care should be taken with significant figures and units. Remember that *every* number read from a device in the laboratory is to be recorded as a piece of raw data.

b. **Calculations.** Here you should present the results of your calculations based on your data, such as determination of solution molarity. When there are several calculations of the same type to be performed, you should set up one "sample calculation" to illustrate the method, and then report only the results of the other, similar, calculations. All sample calculations should include appropriate units for all quantities used in the calculation, as well as for the result. A brief heading or label indicating what it is you are calculating is desirable.

Most students use pocket calculators to perform arithmetic operations after setting up a calculation. If you do the calculations out by hand, you should set aside a page or part of a page for all your arithmetic. Such calculations should appear in the notebook (neatly) but should not be scattered through the report since they tend to obscure the organization.

c. **Unknown.** In experiments where you are asked to characterize one or more unknowns, it may be a good idea to put the data and calculations pertaining to this in a separate section.

d. **Questions.** Some experiments contain questions posed in the sections on treatment of data and on report preparation. These questions should be answered in your report. [The answers to "QUESTIONS" at the end of the write-up which are intended as a guide to your pre-study are generally not put into your report unless your instructor asks you to.]

e. **Error analysis.** You should close each report with an estimate of the reliability of your results. There is further discussion of this matter in the section on "Experimental Precision and Significant Figures."

As a simple example of some of the points of how a notebook record should appear, suppose you were determining the density of a small cube of copper. Suppose you measured the lengths of the edges with a ruler having millimeter markings. This would enable you to estimate distances to the nearest tenth of a millimeter. Let us suppose further that you weighed the cube on a Mettler balance, which gives weights to the nearest tenth of a milligram. Your notebook might look as shown on the next page.

You will note that the raw data are clearly identified and experimental uncertainties for each quantity are given. The method of calculation is indicated, including units for each number. And finally, an error limit, or uncertainty, for the final result is given. In this case, the final answer has three significant digits as a result of the fact that the original length measurements were good to only three significant figures. A detailed discussion of the rules for handling significant figures is given in a later section.

Joan Doe
Experiment 25
April 1, 1984

p.34

Determination of Density of Copper Cube

Cube number 8
Room temp. 24.2° C.

DATA   Length of cube edge   11.3 mm   ± 0.1 mm
                 "   "   2nd   "   11.3 mm   "
                 "   "   3rd   "   11.3 mm   "

[All three edges emanate from one corner
and the intersections appear perpendicular.
This supports, but does not prove, object to be a cube.]

Mass of cube   12.8273 g   ± 0.0001 g

[Since object was metallic solid, weighed
directly on balance pan. Only one cube
reading necessary.]

CAL'N $Density = \dfrac{M}{Vol} = \dfrac{12.8273\,g}{(1.13\,cm)^3} = 8.88996 = 8.89\,g/cm^3$
                                                                not
                                                            significant

ERROR ANAL.   Mass   12.8273 ± 0.0001 $\Rightarrow$ 0.0008 %
                        Length   1.13   ± 0.01 $\Rightarrow$ 0.9 %

$Uncertainty_{Density} = U_M + U_{Vol} = U_M + 3U_{Leng.} = 2.7\% \sim 3\%$

$\therefore Density = 8.89\,g/cm^3 \pm 3\% = \underline{8.89 \pm 0.27\,g/cm^3}$

Figure 1. Sample Notebook Page.

7

Experimental Precision and Significant Figures

Any quantity measured experimentally will have some uncertainty associated with it. This uncertainty cannot be avoided, since it arises from the limitations of the instruments or apparatus being used. Every piece of apparatus naturally has some smallest sub-division that one can read on its measuring scale, and this ultimately determines the precision with which a measurement can be made. Even if an experimenter estimates beyond what can be read directly from the scale, he is limited by the extent of the sub-divisions. For example, a meter stick is usually marked off in millimeters. An experimenter using the meter stick might estimate a length to tenths of millimeters, but there is no way he could use the meter stick to get a value good to hundredths of a millimeter. To do this he would need to use a different (more precise) measuring device.

Precision

Let us consider this limitation imposed by laboratory instruments more carefully. Suppose, for example, several people independently read a thermometer measuring the temperature of a beaker of water before and after adding an additional quantity of hot water to the beaker. Each person recorded the initial and final temperatures, subtracted the measurements, and obtained a value for the temperature rise. Even though they all observed the same thermometer as carefully as they could, they would most likely have different values for the result, since it is necessary to estimate the last figure in reading a thermometer scale.

If the thermometer were graduated in degrees, it is reasonable to expect that a temperature of, say, 25.4°C to be read as between 25.3 and 25.5. We could write this as 25.4 ± 0.1°; the ± 0.1 (read as "plus or minus 0.1") gives the uncertainty in the reading. A similar uncertainty in the final reading (say, 35.8 ± 0.1°) gives us a temperature rise of 10.4 ± 0.2°, where we have added the uncertainties to get an outside (or maximum) uncertainty. The total spread in ΔT is 0.4°C, or a variation of ± 0.2° about the average value. Thus, we might expect the various observers to give temperature rises ranging from 10.2° to 10.6°. In symbols

$$\Delta T = (T_{final} - T_{initial}) \pm \frac{\text{Spread}}{2}$$

$$= 10.4 \pm 0.2°$$

Note that the expected smallest and largest values of ΔT arise from the combinations of the extreme values of the measurements:

$$35.7 - 25.5 = 10.2 \qquad 35.9 - 25.3 = 10.6$$

The spread in values for a measurement indicates the **precision** of the measurement. *Precision refers to the degree of uncertainty in a measurement due to limitations in the apparatus; it gives the range of values which would be obtained by repeated measurements on the identical system.* Precision can be stated in terms of the ± value, but it is often expressed in a relative sense in terms of parts per million (ppm), parts per thousand, or percent (parts per hundred). Thus, in the above example, the precision is

$$\pm\ 0.2° \quad \text{or} \quad \frac{2}{104} = 0.02$$
$$= 2\%$$
$$= 20 \text{ ppt}$$
$$= 2 \times 10^4 \text{ ppm}$$

Here we have used but one significant figure since the original error estimate contains but one significant figure. It is most common to express precision as a percentage. The last two methods of expressing precision (parts per thousand and parts per million) will not be used often in Chemistry 14.

As a second example of the precision of a measurement, consider the determination of the weight of a sample of a solid salt. A single-pan analytical balance of the type used in general chemistry is capable of measuring to the nearest 0.1 milligram (10^{-4} gram). Therefore, we might have the weight of a container-plus-sample of 20.3964 ± 0.0001 g and a weight of the container alone of 18.2142 ± 0.0001 g. This gives a weight of sample, by difference, of 2.1822 g with an uncertainty of $\pm\ 0.0002$ g. The precision of this measurement is

$$\frac{0.0002}{2.1822} \approx \frac{2}{20000} = 0.1 \text{ part per thousand} = 0.01\%$$

Thus, the measurement of the weight of this sample is much more precise than the determination of ΔT in the first example.

Absolute and Relative Errors

In the above cases, the uncertainties have been expressed both on an absolute basis and a relative basis. It will be helpful to give exact definitions for these. By **absolute error** we mean the $\pm$ value of the measurement (or one-half the spread of the values). The absolute error must naturally have the *same units* as the measurement to which it refers. Alternatively, the uncertainty may be expressed on a relative basis. The **relative error** is the ratio of the absolute error to the value of the measurement. Since it is the ratio of two numbers with the same unit, a relative error is *dimensionless*. For the examples given above we have

Measurement	Absolute Error	Relative Error
Temperature $10.4° \pm 0.2°$	$\pm\ 0.2°$	$\frac{0.2}{10.4} = 2\%$
Weight $\quad 2.1822 \pm 0.0002$ g	$\pm\ 0.0002$ g	$\frac{0.0002}{2.1820} = 0.01\%$

You should note that the percent error (or relative error) in a measurement depends upon the *magnitude* of the measurement. Often this means it depends upon the amount of material you are working with. The outside absolute error in a weighing is about 0.0002 g regardless of the total mass. As indicated above, this gives a 0.01% error for a mass of 2 g. But if you weighed out only 0.02 g of a substance the error is still 0.0002 g absolute. This produces a relative error of 1%. The amounts of material used in an experiment are based partly on a goal of reducing errors.

Exact and Inexact Numbers

The experimental uncertainty in measurements such as the temperature rise and the weight discussed above means that the values obtained are inexact numbers. Indeed, any experimental measurement—volume of a gas, voltage of a cell, number of calories released in a reaction, etc.—will be inexact since it cannot be known with infinite precision. The preceding discussion of precision and the later discussion of significant digits apply to these **inexact numbers.**

By contrast, there are certain numbers in chemistry which are exact, with no uncertainty. Usually **exact numbers** are the result of counting discrete objects, e.g., counting the number of students in a class. Barring a blunder in doing the counting, the resulting integer tells exactly how many students are present. If 25 students are counted as being present, this means exactly 25—not 25 ± 0.1 or 25 ± 0.001, but precisely 25. The number can be thought of as having *an infinite number of zeros* after the decimal.

In chemistry, exact numbers occur frequently in calculations, and often arise from counting. For example the formula, H_2O, means that there are exactly two atoms of hydrogen to one atom of oxygen.

$$\frac{\text{number of hydrogen atoms}}{\text{number of oxygen atoms}} = 2 \text{ or } 2.000000000 \ldots \text{ (exact)}$$

The precision of any calculated result using this ratio will not be limited by the ratio. Another common source of exact numbers is definitions. A centimeter is defined as one-hundredth of a meter, or 1 m = 100 cm. This is an exact ratio, and likewise a calculation involving the unit conversion will not be limited by the precision of the ratio.

Significant Digits

In the examples above, the measurements all had an uncertainty in the last digit. They conform to the general rule, commonly used by scientists, that unless indicated to the contrary digits written in a measurement are considered to be known with certainty except for the last digit, which is assumed to be somewhat in doubt. It is common practice to omit the ± value, with the understanding that the number reflects the above rule; that is, only the last digit is uncertain. The digits which represent the measurement, including the last "somewhat doubtful" digit, are called **significant digits** (also sometimes referred to as significant figures).

It is useful to describe a measurement by stating the *number* of significant digits which it contains. Such a description gives a general idea of the precision, since *the larger the number of significant digits, the more precise the measurement:*

9.4 g   two significant digits, $\backsim$ 1% precision

843.6 g   four significant digits, $\backsim$ 0.01% precision

The second measurement containing the larger number of significant digits is obviously more precise.

In order to apply the principle covered in the preceding paragraph, it is necessary to know how to count the number of significant digits in a number. This ability will also be important in considering how calculated results should be expressed. The rules which govern the counting of significant digits are simple and straight forward:

1. *All digits other than zero are always significant.*
2. *Zeros are significant except when they only serve to locate a decimal.*

Whether a particular zero should be considered significant sometimes causes confusion. The following examples will help make the situation clear.

40.3 g	*three* significant digits; the zero in the units' place has meaning, since the measurement has been made to tenths.
32.90 ml	*four* significant digits; there is no reason to write the zero in the hundredths place unless the measurement had been made to this precision.
0.032 g	*two* significant digits; the zeros to the left before the first non-zero digit merely serve to place the decimal. The number could better be written as 3.2×10^{-2}, which clearly shows two significant digits.
450 ml	*two or three* significant digits; the zero may be significant *or* may serve only to place the decimal. If the volume were measured to ± 1 ml, then the zero is significant; if the volume were measured roughly as being between 440 and 460 ml, then the zero is meaningless.

The ambiguity of cases like the last one can be avoided by the use of exponential notation for expressing quantities. For example, a distance of 20500 mm can be written with exponential notation to reveal clearly the precision of the measurement:

2.05×10^4 mm	three significant digits.
2.050×10^4 mm	four significant digits.
2.0500×10^4 mm	five significant digits.

Many students develop a careless tendency to drop final zeros to the right of the decimal even when they have meaning. *You should always avoid this tendency.* For example, suppose you use a buret for delivering a volume of liquid and collect the following data:

Final buret reading	28.53 ml
Initial buret reading	13.53 ml
	15.00 ml (not 15 ml)

The difference in these two readings gives the volume of liquid delivered, which should be reported as 15.00 ml. It is incorrect to shorten this number to 15 ml, because the precision of the measurement means that the hundredths place contains a significant digit, in this case a zero.

Significant Digits in a Calculated Quantity

The data obtained in measurements generally are used to calculate some derived quantity. For example, from the measurement of the mass and of the volume of a liquid sample, you could calculate its density, a **derived quantity**. The precision of the derived quantity is dependent upon the precision of the measurements which are used in its calculation. It is important, therefore, that the result be expressed in such a way as to reflect this dependence. One method to do this is to use the proper number of significant figures in the calculated result.

There are two general ways in which numbers are used in calculations, namely, *addition* (or subtraction, which is the addition of a negative) and *multiplication* (or its complement, division). Accordingly, two rules are needed for these two different types of calculations:

I. When numbers are added (or subtracted), *the result contains the same number of decimal places as does the term with the fewest number of decimal places.*

11

For example, suppose 0.256 g of NaCl and 5.42 g of KBr are both dissolved in 87.1 g of water. What is the total mass of the solution?

$$
\begin{array}{r}
87.1 \\
0.256 \\
\underline{5.42} \\
92.776 \text{ g}
\end{array}
$$

The first reaction is to say the mass is 92.776 g. This is incorrect, however, since it would imply that the second and third decimal places are known. But the mass of water is uncertain in the tenths' place:

$$
\begin{array}{r}
87.\underline{1} \\
0.25\underline{6} \\
\underline{5.4\underline{2}} \\
92.\underline{776} \text{ g} \Rightarrow 92.\underline{8} \text{ g}
\end{array}
$$

Therefore, the answer must be uncertain in the tenths' place, and the last two digits in the sum cannot be included. The answer must be rounded off as indicated.

This rule means that in addition and subtraction the answer contains no digit beyond the place in which any one of the contributing terms has a doubtful digit.

II. When numbers are multiplied (or divided), *the result contains the same number of significant digits as the factor with the fewest significant digits.*

$$11.8859 \times 7.5 = 89.14425 \Rightarrow 89$$

$$\uparrow$$

2 sig. dig.   Round off to 2 significant digits.

The correctness of this rule becomes evident when the meaning of 7.5 is recalled, namely 7.5 ± 0.1. If 11.8859 is multiplied by 7.4 and by 7.6, the answers will cover the range corresponding to the uncertainty in 7.5:

$$11.8859 \times 7.4 = 87.95566 \Rightarrow 88$$

$$11.8859 \times 7.6 = 90.33280 \Rightarrow 90$$

Clearly the digit in the units' place is uncertain, and thus all of the digits to the right of the decimal are completely unknown. The result of the multiplication must be expressed as 89 ± 1, or simply 89.

One of the most abused facets of the use of significant digits is the inclusion of many more (in)significant digits than are justified. Obviously in the above case it is incorrect to give the answer as 89.14425, although this is what you might read from a calculator. You must resist the tendency to copy *blindly* the answer a calculator gives.

In a reverse way, you must sometimes remember to add to what a calculator produces as its answer. For the following multiplication, a calculator will give the answer shown.

$$15.625 \times 0.0128 = 0.2$$

This is mathematically correct, since calculators suppress terminal zeros. But the correct answer, taking into account significant digits, is 0.200. You must be alert to situations like this and include the correct number of digits.

Rounding Off Numbers

In the preceding section the concept of rounding off a number to reduce the number of digits was used. This procedure must be examined in more detail. The general rules are the following. If the *first digit* (i.e., the leftmost) *to be dropped is less than 5,* the last retained digit is *left unchanged.* If the *first dropped digit is greater than 5, or is 5 followed by any non-zero digits,* the last retained digit is *increased by one.* The special case when the *first digit to be dropped is exactly 5* is handled by rounding so the resulting number is *even.* This rule prevents cumulative errors which might arise in a lengthy calculation if numbers ending in 5 were always rounded to the higher value.

The rules for rounding are illustrated in the following examples of numbers rounded to four digits.

13.7849	$\Rightarrow$ 13.78	4 less than 5; round *low*
13.786	$\Rightarrow$ 13.79	6 greater than 5; round *high*
94.7350	$\Rightarrow$ 94.74	50 exactly; round *even*
94.785	$\Rightarrow$ 94.78	5 exactly; round *even*
26.325001	$\Rightarrow$ 26.33	5001 greater than 5; round *high*

Intermediate results which are used in further calculations are subject to special consideration. If each intermediate result is rounded to the proper number of significant digits according to the earlier rules, the final result may be slightly in error because of a cumulative effect. To avoid this situation, a common practice is to retain one more digit in intermediate answers than is justified by the data. This extra digit is sometimes distinguished by writing it as a subscript.

$$7.314 \text{ ml} \times 0.5325 \text{ g/ml} = 3.894705 \text{ g}$$

As a final result   3.895 g

As an intermediate result 3.894_7 g

Eventually, any final result based on using this quantity must be expressed with no more than four significant digits.

Errors and Their Analysis

In Chemistry 14 you will be required to determine and report the numerical value of certain physical quantities such as the heat of formation of solid NH_4Cl or the percent composition of a complex salt. These determinations require that some specific measurements be made using laboratory equipment of different precision. As was discussed in the section on "Precision," your results, and in fact the results obtained by any experimenter, will always be subject to uncertainties brought about by various possible error sources. The experimenter must be aware of these error sources and must also give an estimate of the uncertainty in his reported results. A reported result for which the uncertainty is not known is of little value to either the experimenter or the reader.

Types of Errors

It is convenient to consider the possible errors in a measurement as classified under three broad headings.

A. Random errors (indeterminate errors)

These are the uncertainties that arise because of the limitations of apparatus and instruments. Each given measurement will have an error, but the error does not have a definite assignable value (it is **indeterminate**). The magnitude of the error fluctuates in a random manner within the limits, or spread for the measurement. This is the kind of error which was discussed in the section on "Experimental Precision" (p. 8). Random errors will be the primary subject of your error analysis of an experiment.

B. Personal errors

An experimenter may be careless, ignorant, or make "mistakes." For example, he might incorrectly record a 42 or 24, or he might use a piece of apparatus incorrectly. If personal errors are known to exist, they can usually be eliminated through the exercise of proper care on the part of the experimenter. [Personal errors are sometimes included under the next heading as one kind of determinate error.]

C. Systematic errors (determinate errors)

These are errors introduced by some defect in a procedure, or by an improperly calibrated instrument. A **determinate error** has a definite value which can (in principle, if not in practice) be measured and accounted for.

As an example of a systematic error, consider a precise thermometer which can be read to ± 0.02°C. This is ten times more precise than the thermometer mentioned on page 8, which could be read to only ± 0.2°C. Now suppose that this precise thermometer was manufactured with a calibration defect, the defect being that the temperature one reads from the scale is exactly 0.5°C lower than the actual temperature. Any temperature measured with this thermometer would be consistently lower than the true temperature. Thus it would be in error despite the fact that measurements could be made precisely (to ± 0.02°C). The temperatures would be inaccurate because of a systematic error—an error which is present in every measurement because of a non-valid assumption made by the experimenter.

A systematic error causes all measurements in a series to be off in a similar manner. A systematic error could arise because a chemical was impure, a solution concentration was wrong, a balance was incorrectly calibrated, etc. Systematic errors, if detected, can be corrected for; unfortunately, they are sometimes difficult to detect.

Accuracy

In the preceding paragraph, we noted that highly precise measurements could be incorrect because of systematic errors which affect all of the measurements in a set in the same way. This brings us to the difference between precision and accuracy.

The **accuracy** of any measurement is the degree to which the measurement matches the "true" value of the measured quantity. (Contrast this with the meaning of precision, which refers to the agreement among repetitive experimental measurements.) To determine the accuracy, we must know the true value. But the true value must initially be obtained from experimental measurements which may contain systematic errors. We are then confronted with the basic question, *"When do we accept an experimental result as a true value?"*

The answer to this question is the following. If independent experimenters working in different laboratories and using entirely independent methods arrive at very similar and precise results, we assume that their results are free from systematic error and hence represent the true value of the measured quantity. For example, the atomic weights which are accepted as "true" are a result of applying this criterion. In your laboratory work, the true values are ones that are based on results generally accepted by the scientific community as a consequence of multiple verification and rechecking.

NOTE ON USAGE

The word "error" strictly refers to the difference between an experimental value and the true value for the property. It thus refers to the accuracy of the measurement and not to its precision. For reference to precision, treatments of error analysis at a more advanced level use the term "deviation" to refer to the spread in experimental values in repeated measurements made under identical conditions. Nevertheless, it is common practice to use terms such as "random error" when referring to the precision of measurement. This manual retains that usage.

By definition of the meaning of precision, there is no such value as the precision of a single measurement. It is proper, however, to calculate the uncertainty of a single measurement. This is the quantity we estimate from the readability of a scale, e.g., of a balance or a buret, and which we use in expressing the "random errors" in our work.

Random Errors in Derived Quantities

Typically in an experiment you will measure several properties and then use known relationships to calculate some derived result. You will also need to assign a ± uncertainty to the final calculated result. In order to do this, it will be necessary to know how the uncertainties in the measured properties contribute to the answer. This section presents the rules for evaluating the error in a mathematically derived quantity.

Let us consider some quantity Q which is based on various independent quantities $X, Y, Z. \ldots$

$$Q = f(X, Y, Z \ldots)$$

Each measured quantity has a random uncertainty $\Delta X, \Delta Y, \ldots \ldots$. These errors will cause a random error ΔQ in Q, an error that we must evaluate. The value of ΔQ depends upon the functional form of Q. We shall give without proof rules for these relations.

The two most commonly used functional forms for Q are the following:

I. Q is a sum

$$Q = aX + bY \qquad a \text{ and } b \text{ are constants}$$
$$\Delta Q = (a\Delta X) + (b\Delta Y)$$
$$\text{if } a = b = 1, \Delta Q = \Delta X + \Delta Y.$$

[This expression was introduced intuitively on page 8.]

II. Q is a product

$$Q = kXY \qquad k \text{ is a constant}$$
$$\frac{\Delta Q}{Q} = \frac{\Delta X}{X} + \frac{\Delta Y}{Y}*$$

It will perhaps be helpful to restate the above rules in words.

Absolute uncertainty of a sum = sum of the absolute uncertainties of the terms
Relative uncertainty of a product = sum of the relative uncertainties of the factors

Other useful rules are the following.

1. $Q = k\,X^m$ $\dfrac{\Delta Q}{Q} = m\left(\dfrac{\Delta X}{X}\right)$

2. $Q = \dfrac{k\,X^m\,Y^n}{Z^p}$ $\dfrac{\Delta Q}{Q} = \left| m\left(\dfrac{\Delta X}{X}\right)\right| + \left| n\left(\dfrac{\Delta Y}{Y}\right)\right| + \left| p\left(\dfrac{\Delta Z}{Z}\right)\right|$

3. $Q = \ln X$ $\Delta Q = \dfrac{\Delta X}{X}$

4. $Q = \log_{10} X$ $\Delta Q = \dfrac{1}{2.303}\dfrac{\Delta X}{X}$

As an illustration of the general procedures to be followed in applying these rules, consider the following example.

A solution is prepared by dissolving 0.0200 moles of NaCl in enough water to give 100 ml of solution. The NaCl was measured to ± 0.0004 moles and the solution volume was measured to ± 3 ml. What is the molarity of the solution (including its uncertainty)?

*A justification for this expression may be given as follows.

$$(Q \pm \Delta Q) = k\,(X \pm \Delta X)(Y \pm \Delta Y)$$
$$= k\,[XY + Y\Delta X + X\Delta Y + \Delta X\Delta Y]$$

Since in general ΔX and ΔY are small, then $(\Delta X\Delta Y) \ll X\Delta Y$, and $\Delta X\Delta Y$ can be neglected. We can thus write

$$Q \pm \Delta Q = kXY + k\,[Y\Delta X + X\Delta Y]$$
$$\Delta Q = k\,[Y\Delta X + X\Delta Y]$$
$$\frac{\Delta Q}{Q} = \frac{k\,[Y\Delta X + X\Delta Y]}{kXY} = \frac{\Delta X}{X} + \frac{\Delta Y}{Y}$$

$$\text{MOLARITY} \equiv \frac{\text{moles solute}}{\text{liters solution}} \qquad M_{NaCl} = \frac{0.0200 \text{ mole}}{0.100\ell} = 0.200 \text{ M}$$

This is the calculated value, to which we must now assign an uncertainty. Applying the above general rule for a product (quotient), we have

$$\text{\% error in molarity} = \text{\% error in moles} + \text{\% error in volume}$$

$$\left(\text{or } \frac{\Delta M}{M} = \frac{\Delta \text{ mole}}{\text{mole}} + \frac{\Delta \text{ liter}}{\text{liter}} \right)$$

Thus, to get $\pm M$, we must calculate the percent errors in the measured quantities.

	Value	Uncertainty	% Error	
moles:	0.0200 moles	± 0.0004 moles	2%	$\left(\frac{0.0004}{0.0200} \times 100 \right)$
volume:	100 ml	± 3 ml	3%	(Simply, 3 in 100)

Therefore, the error in M is 5% (sum of 2% + 3%)

$$\text{Concentration} = 0.200 \, M \pm 5\%$$
$$= (0.20 \quad \pm \quad 0.01) \, M$$

Note in the final value, only two significant figures are justified, since 5% of 0.200 = 0.01, not 0.010. Since the uncertainty is in the hundredths place, you are not justified in expressing M to better than 0.20!

ADDENDUM: When quantities are combined by adding or subtracting, the absolute errors in each value are added. An example is the weight of a sample obtained by weighing a container with and without the sample. The uncertainty in the sample weight is twice the precision of a single weighing. Note that anytime two quantities are added (subtracted) they must have the same units.

Use for Graphs for Derived Quantities

Sometimes a derived quantity is obtained from the slope of a straight line graph. In such cases, the uncertainty in the derived quantity can be approximated graphically.

For example, consider the relation $y = ax + b$ where y and x are measurable quantities and a is the quantity to be derived. A plot of y vs. x should yield a straight line of slope a. The uncertainty in this slope can be approximated as follows:

1. Plot y vs. x and reject any poor data.
2. Draw the best straight line through your experimental points; this line must pass through, or close to, the largest possible number of points.
3. Measure the slope a' of this line. [For details, see pages 29-30.]
4. Draw two additional dashed lines representing what reasonably appear to be the maximum and minimum slopes.
5. The difference between the slopes of the two dashed lines represents the spread S in the value of the slope a. This spread is twice the expected maximum error in a.

6. Report your result as

$$a = a' \pm \frac{S}{2}.$$

This graphical method will be used in Topic 4.

General Approach to Error Analysis

For each experiment you will be expected to include an analysis of the reliability of your results. This can most easily be accomplished by following the general procedure outlined below.

A.   1. For each measured quantity, estimate its uncertainty.
        2. Using the rules for the errors in derived quantities, calculate the uncertainty in the answer. Designate this value U_m and indicate it with your results.*
        3. Indicate the measurement that contributes the most to the uncertainty U_m and, if possible, suggest simple changes in the experimental procedure that could lead to a lower uncertainty (i.e., higher precision results).

B.   If the accepted true value of the quantity you determined is known, then in addition to the operations mentioned in A above, you should do the following.

    1. Calculate the actual percent error E as

$$E = \left| \frac{\text{True value} - \text{Your value}}{\text{True Value}} \right| \times 100.$$

    2. Compare E with U_m.

       a. If $E < U_m$, then you are justified in assuming that the net systematic error in your results is probably less than U_m.
       b. If E is appreciably larger than U_m, then either you committed a careless mistake, or the systematic error is large, or both. Checking your measurements and calculations will in general quickly tell you whether a careless mistake has been committed. If no such mistake is detected, you should consider repeating the experiment. If this is not possible, you should attempt to indicate the most probable reason for your large systematic error.

The two cases, where $E \leqslant U_m$ and $E > U_m$, will be considered in the following examples.

Example of Error Analysis

Suppose you were to determine the density of a cylindrical metal rod by measuring its dimensions to get the volume and by weighing it on a triple-beam balance to get its mass. Assume the ruler you used could be read to ± 0.5 mm and the balance could be read to ± 0.1 g. Your notebook record, with the error analysis, might look as shown in Figure 2. (The numbers 1, 2, and 3 have been added to key the sheet to the steps outlined above. Normally they would not appear in your report.)

---

*U_m is a maximum expected uncertainty due to random errors, since it is calculated on the assumption that the worse happens, namely, that the maximum error occurs simultaneously in all measurements. Actually, random errors tend to cancel each other partially, thus leading to a net random error which is less than U_m.

Determination of the density of a metal rod

DATA

$$
\begin{aligned}
&\text{Mass of rod} \quad (M) \quad 34.6\,g \quad (\pm 0.1) \\
&\text{Length of rod} \quad (L) \quad 10.04\,cm \quad (\pm 0.05) \\
&\text{Diameter of rod} \quad (D) \quad 1.25\,cm \quad ''
\end{aligned}
$$

CALCULATIONS

$$\text{Density} = \frac{M}{V} = \frac{M}{A \times L} \qquad A = \text{area (cross-sectional)} = \pi\left(\frac{D}{2}\right)^2$$

$$= \frac{34.6\,g}{(3.14)\left(\frac{1.25\,cm}{2}\right)^2 (10.04\,cm)} = 2.81\,g/cm^3$$

ERROR ANALYSIS

1

	Magnitude	Absolute error	Relative error
M	34.6	0.1	0.3%
L	10.04	0.05	0.5%
D	1.25	0.05	4 %

2

$$\text{Density} \propto \frac{M}{L \times D^2} \qquad U_{Den.} = U_M + U_L + 2U_D$$

$$= 0.3 + 0.5 + 2 \times 4 = 8.8 \sim 9\%$$

$$\text{Density} = 2.81 \pm 9\% = \underline{2.81 \pm 0.25\,g/cm^3}$$

3   The measurement of the diameter contributes most to the uncertainty U_m in density. If the diameter could be measured to one more significant figure, its contribution could be reduced by a factor of ten (from 8% to 0.8%). Maybe use a vernier caliper or a micrometer. Even with a more precise diameter measurement, it still would make the major contribution. This is partly because the diameter is relatively small compared to the length. Its contribution is also doubled, since density depends upon D^2.

Figure 2. Sample Notebook Page with Error Analysis.

As an extension of the above hypothetical experiment, suppose that the rod is believed to be made of aluminum. The true density of aluminum is 2.70 g cm^{-3}. Now your error analysis, in addition to Part A, should have a discussion relating to Part B of the outline on page 18. Your notebook record of this part might then appear like Figure 3.

Actual error in density

$$E = \left| \frac{2.70 - 2.81}{2.70} \right| \times 100 = 4.1\%$$

Since $E < U_m$ (4.1 < 9), I can assume the net systematic error is less than 9%. My result is correct within the maximum random experimental error expected.

Figure 3. Sample Discussion, E < U$_m$.

Finally, suppose that using the same equipment as before you had got an experimental density of 3.20 g cm^{-3} (rather than 2.81 found before) for this supposedly aluminum rod. This corresponds to the actual error, E, being greater than U_m. Your analysis might then appear like Figure 4.

Actual error in density

$$E = \left| \frac{2.70 - 3.20}{2.70} \right| \times 100 = 19\%$$

This actual error is much larger than the expected maximum uncertainty, U_m. U_m = 9%. I have checked all my data and calculations and can find no mistake. Possible source of discrepancy —

1. Maybe the rod is not uniform. If the diameter I measured is less than the average diameter, then the volume calc'l would be too small and I would get too large a density.
2. Maybe the rod isn't pure Al, but rather is made of some more dense metal.

Figure 4. Sample Discussion, E > U$_m$.

Although the analysis given in the above example is not very elaborate, it adds great significance to the reported results. For, as a result of this analysis both the experimenter and the report reader know (a) the degree of uncertainty in the results; (b) which measurement contributes the most to this uncertainty; (c) how the precision can be improved; and (d) if the accepted true value is known, how serious is systematic error.

Clearly, a consideration of the reliability of a result must be a standard step in any scientific investigation. In this area, the goal of Chemistry 14/15 for the student is two-fold:

a. To supply practice, and thus develop competency, in expressing uncertainties in measurements and in answers both on an absolute and on a percentage basis,

b. To develop an attitude of recognition of the existence of an uncertainty in any experimental measurement.

It is further expected that the student will develop the ability to recognize which measurement in an experiment may be limiting because of an inherent lack of precision. When this recognition is achieved, the student will come to understand the futility of being extremely careful with some measurements or conditions in an experiment when some other very imprecise datum is limiting. No effort was made to introduce the concepts of standard deviation and error statistics. These concepts will be taken up in more advanced courses.

After reading these seemingly long notes on precision and error analysis, you may feel that error analysis will be a very difficult task for you. If so, you should *reread* these notes carefully and perhaps more than once. The next section contains a summary of terminology with examples that should aid your understanding. The apparent difficulty you perceive is probably because a consideration of errors is something new to you. You will find that your error analysis will improve very quickly with experience and will be a valuable addition to your abilities as you proceed in other courses requiring experimental measurements.

Summary of Error Terminology

1. Random Error

   In any quantitative scientific measurement, there exists an uncertainty or random error. This uncertainty, resulting from limitations of the instruments being used, cannot be avoided.

   A. Absolute Error

      The absolute error is defined as the uncertainty in a measurement. The absolute error has the same units (grams, °C, liters, etc.) as the measurement in question.

      Example: A Mettler balance can be read to the nearest tenth of a milligram (10^{-4} gram). What is the absolute error in the measured quantity 12.3122 g?

      ANS. 0.0001 g (10^{-4} g). The above value would be expressed as 12.3122 ± 0.0001 g.

      Example: 50 ml of liquid are in a graduated cylinder which can be read to the nearest 1 ml. What range of volumes is possible for the liquid?

      ANS. The absolute error is 1 ml. Express the measurement as 50 ± 1 ml. The volume of the liquid could be as little as 49 ml or as much as 51 ml.

   B. Relative Error

      The relative error is the ratio of the absolute error to the measurement. It may be expressed as a number or as a percentage.

      Example: A measured quantity of 46.7775 g is uncertain by 0.0002 g. What is the relative error?

      ANS. $\dfrac{0.0002 \text{ g}}{46.7775 \text{ g}}$ = 0.000004 = 0.0004% = 0.004 ppt = 4 ppm.

      The last two methods of expressing the relative error (parts per thousand and parts per million) are not used in this course as often as the first two.

      Note that the relative error is unitless, since the units always will cancel.

      Example: What is the relative error (%) in a 60.0 ml measurement if the absolute error is 0.5 ml?

      ANS. $\dfrac{0.5 \text{ ml}}{60.0 \text{ ml}}$ × 100% = 0.8%

      Example: A mass of 15.87 g is weighed with a precision of 0.5%. What is the range of possible values of the mass?

      ANS. 15.87 × 0.005 = 0.08 g absolute error. Mass is 15.87 ± 0.08, or a range of 15.79 − 15.95.

      Be careful not to mix decimal and percent expressions. For example, 0.01 ≠ 0.01%.

      Do not add the absolute error to the denominator (measurement).

2. Random Error in Calculated Values

   A. Error in a Sum

The absolute error in a sum is found by adding the absolute errors of the numbers contributing to the sum.

Example: A beaker on a Mettler balance (uncertainty 0.0001 g) weighs 50.0885 g when it is empty. A solid is put into the beaker; the combined weight is 67.9663 g. What is the weight of the solid? What is the relative error?

ANS.

$$\begin{array}{r} 67.9663 \pm 0.0001 \text{ g} \\ -50.0885 \pm 0.0001 \text{ g} \\ \hline 17.8778 \pm 0.0002 \text{ g} \end{array}$$

$$\frac{0.0002 \text{ g}}{17.8778 \text{ g}} \times 100\% = 0.001\%$$

Example: An initial temperature is $21.0 \pm 0.5°C$ and the final temperature is $78.0 \pm 0.5°C$. What is the value of the temperature rise and the relative error in that value?

ANS. $78.0 \pm 0.5°C - (21.0 \pm 0.05°C) = 57 \pm 1°C$

$$\frac{1°C}{57°C} \times 100\% = 2\%$$

When there is an initial and a final reading on an instrument, the absolute error in the measurement will be twice the instrumental uncertainty.

Do not add relative errors here.

B. Error in a Product

Add the relative errors of the measurements contributing to the product to obtain the relative error in the product.

Example: 50 ± 1 ml of a 0.10 ± 0.005 M solution contains how many moles? What is the relative error?

ANS. 0.050 liter $\times$ 0.10 mole/liter = 0.0050 mole.

$$\frac{1 \text{ ml}}{50 \text{ ml}} = 2\% \text{ error} \qquad \frac{0.005}{0.10 \text{ M}} = 5\% \text{ error.}$$

The relative error in the answer is 2% + 5% = 7%. This is the maximum expected relative error, symbolized U_m on page 18.

3. Systematic Error

A systematic error occurs when there is something "wrong" with the experiment, i.e., an error not due to instrumental uncertainty. Systematic errors include leaks, contamination, improperly adjusted (calibrated) instruments, neglecting a quantity that actually is not negligible, etc.

A. True Error

The true error is defined as follows:

$$\left| \frac{\text{Experimental value} - \text{Actual value}}{\text{Actual Value}} \right| \times 100\%$$

Example: A molecular weight is found experimentally to be 122.10 g/mole. It is known to be 120.64 g/mole. What is the error?

ANS. $\dfrac{122.10 \text{ (g/mole)} - 120.64 \text{ (g/mole)}}{120.64 \text{ (g/mole)}} \times 100\% = 1.21\%$

B. Detection of Systematic Error

If the true error exceeds the (relative) random error, a systematic error is indicated, because if instrumental error only were involved, the true error would be less than or equal to the random error.

Example: The value of U_m in a result is 4%, but the experimental error is 10%. This indicates the presence of either a systematic error or perhaps a blunder.

Making and Interpreting Line Graphs

Introduction

In an experimental science like chemistry we are frequently seeking generalizations from collected data. Typically, we proceed by setting the value of one property, called the **independent variable**. We then measure the value of another property, the magnitude of which depends upon that of the independent variable. This second measured property is called the **dependent variable**. Thus, we obtain data consisting of pairs of numbers, one for the independent variable and the other for the dependent variable. Our goal is to recognize the basic relationship between the variables.

Data could be presented simply in a table, but the numbers generally do not give an immediate idea of the nature of the relation between the properties. Often a presentation in a graph is more useful in discovering the relation. The graph is merely a pictorial representation of numerical quantities so that they can be understood more readily. We shall be interested in line graphs which represent data points.

As an example of the preparation of a plot representing pairs of data points, let us consider the relation between the temperature and the pressure of a sample of gas confined in a closed container of constant volume. The independent variable, temperature, is set at different values and the corresponding pressure exerted by the gas is measured to give data like the following:

DATA	Temperature, °C	100	140	180	250	300
	Pressure, torr	150	166	182	211	231

A plot of these data is developed in Figure 5. By convention, the axes are drawn on the left and at the bottom of the graph. The intersection of the axes here is taken as the point $(T = 0, P = 0)$, and scales are added along the two axes. The method for selecting the size of the scale units along the two axes will be discussed later. Note, however, that in this case the scale units are not the same along the two axes (a given length on the x-axis represents twice as many units as on the y-axis).

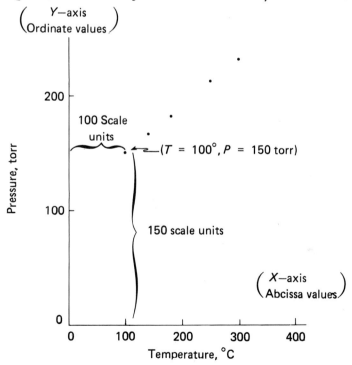

Figure 5. Development of a Graph (a) Plot of Points.

25

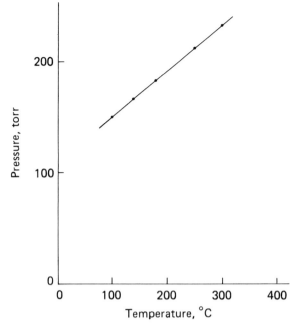

Figure 5. Development of a Graph (b) Line to Represent Data.

Each data point is represented by placing a dot on the graph above its x-axis value (or below, if the y-value is negative) and at a level corresponding to its y-axis value. Thus, the point ($T = 100°$, $P = 150$ torr) is plotted as shown. Figure 5(a) shows the complete set of data points plotted to give the graph. After all the points have been plotted, a smooth curve is drawn which best represents the points. In this case a straight line obviously "fits" the points, as shown in Figure 5(b).

Developing a Graph

The following steps are the ones to follow in making a graph of experimental data.

1. **Draw the axes.** Use a ruler to make straight, neat lines. Often the axes are not located along the edges of the ruled area, but are moved in one or more major divisions so as to allow room along the bottom and the left-hand side for labeling the axes. (This step must be done in conjunction with step 3.)
2. **Decide which data goes on each axis.** The independent variable is normally plotted as the abscissa (x-axis) while the dependent variable is plotted as the ordinate.
3. **Decide on the scales to be used on each axis.** This is the most critical step in producing a satisfactory graph. It involves determining answers to such questions as: What scale size should be used on each axis? Should the scale on each axis start at zero and run up to the highest value to be plotted?
   The general rule is to *plot data points so that they spread out to fill the available graph area as completely as possible.* The resulting large graph is both easier to plot and easier to use after it is drawn. This result is achieved by matching the range of values represented by the scale on a given axis to the range of the data to be plotted on the axis.
   An example will be helpful in explaining the application of this principle. Consider the problem of plotting the following data:

x	45	50	55	60	65
y	31	32	33	34	35

If each axis started at zero and used the same length to represent a scale unit, the result would be Figure 6(a). This plot obviously does not satisfy our general rule of utilizing all the available paper, since much space is wasted by starting the numbering on each axis at zero. Clearly it would be more appropriate to start the numbering on each axis near the smallest data value for that axis. When this is done, the graph of Figure 6(b) results. This graph is described by saying that the *zero has been suppressed* on each axis. Figure 6(b) is an improvement, but it is still highly compressed along the y-axis because the same interval of length has been retained to represent one unit. A further improvement can be made by using a larger distance on the y-axis to represent one unit. The graph in Figure 6(c) illustrates this modification. Thus, *the points are spread out as much as possible by selecting a scale unit length independently for each axis and by suppressing the zero on one or both axes, as necessary.*

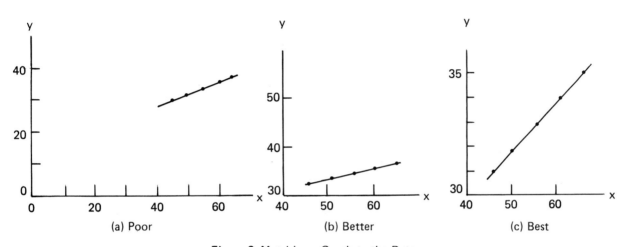

(a) Poor   (b) Better   (c) Best

Figure 6. Matching a Graph to the Data

The choice of scale unit length sometimes must be a compromise between spreading out the points as much as possible and making the plotting as easy as possible. For example, if your graph paper had 20 major intervals along the axis on which you needed to plot values ranging from 60 to 240, you might be tempted to select the scale factor as

$$\frac{(240 - 60) \text{ units}}{20 \text{ intervals}} = 9 \text{ units/intervals}$$

However, this selection would prove very inconvenient because it is not easy to divide an interval into nine equal parts for plotting or for reading values from the graph. An obvious choice would be to assign 10 units per interval. In general, you should avoid taking an interval on the graph paper to be 3, 6, 7, or 9 units. The best choice are those with 1, 2, or 5 (or 0.1, 0.2, 0.5, 10, 20, 50, etc.) units per interval.

The problem of fitting the data to the available graph space clearly involves considering the values to be plotted on both the x-axis and the y-axis. Often a better graph results if the graph is rotated 90° on the paper, placing the x-axis along the long dimension of the paper. The choice will depend upon the range of values for the independent and dependent variables, and upon the number of major intervals available along each dimension of the paper. If the x-axis is placed on the long dimension of a sheet of graph paper, it should appear at the *right* edge, with the origin in the *lower right corner*, when the sheet is included in a report.

4. **Number and label the axes.** The scale chosen is indicated by numbering some of the major division lines. Not every line is numbered because of the lack of space; instead, some convenient multiple is chosen and only these intervals are identified. The numbers should be printed neatly outside the axes. It is good practice to draw short marker lines inside the axes at the numbered intervals.

   Each axis should be labelled clearly to show the property which has been plotted and the units being used. An example is given in Figure 5 where the x-axis is labelled "Temperature, °C" and the y-axis is labelled "Pressure, torr." Note that the unit is *not* repeated with each individual number along the axis.

5. **Plot the data and draw the curve.** The data are plotted as indicated earlier. If data from different sources or different runs are to be included in one plot, they should be distinguished by the use of different marks, for example, closed circles and open circles. The best smooth curve (or straight line, if appropriate) is drawn through the points. The line should *not* obscure the data points. In the case of most experimental data, the uncertainties in the values will result in some scatter about the curve. The curve should be drawn so that the data are "averaged" by placing the line with about as many points on one side of the line as on the other. Do *not* draw a broken line by connecting directly from point to point.

6. **Give a caption to the graph.** Label the graph with a legend which describe its contents, e.g., "The Pressure of a Gas as a Function of Temperature." Include a key if this is needed, as for example, if more than one representation is used for data points.

Straight-line Graphs

Many functional relations in chemistry can be expressed by straight-line graphs. This is fortunate, since a straight line is easy to draw and can be extended, or extrapolated, accurately to give values for points not determined experimentally. An example of extrapolation is shown in Figure 7, which is a replot of Figure 5 with the zero suppressed on the y-axis.

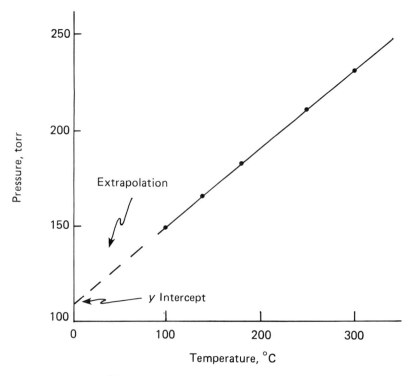

Figure 7. Extrapolation of a Linear Graph.

28

The graph of Figure 7, like any straight-line graph, can be represented by the general equation for a straight line

$$y = mx + b$$

where x and y are variables, while m and b are constants with the following significance:

$$m = \text{slope of line}$$

$$b = y\text{-intercept (value of } y \text{ when } x = 0)$$

The line of Figure 7 can be expressed in the above form if values for m and b are calculated.

$$m = \text{slope} = \frac{\Delta y}{\Delta x} = \frac{\Delta P}{\Delta T}$$

$$= \frac{231 - 150}{300 - 100} = \frac{81}{200} = 0.405 \text{ torr } (°C)^{-1}$$

$$b = y\text{-intercept} = 110 \text{ torr (by extrapolation)}$$

Thus, the pressure-temperature data of Figure 7 are expressed by the equation

$$P = 0.405 \frac{\text{torr}}{°C} \ T \ (°C) + 110 \text{ torr}$$

The slope for a straight line must be a constant. However, when experimental data like the above are used, various pairs of data may give different slopes because the points are not all exactly on the line (experimental uncertainties). For example, the first and fourth points from the above set yield a slightly higher value for the slope

$$\frac{\Delta P}{\Delta T} = \frac{211 - 150}{250 - 100} = \frac{61}{150} = 0.407 \text{ torr } (°C)^{-1}$$

Indeed, the value of the slope of a line should be obtained from two points *on the line*, rather than from two data points. Although the line is the best representation of the entire set of data, the individual data points probably scatter on either side of the line and thus any two will yield a slightly (or markedly) erroneous value for the slope. In the example above, the two extreme data points were taken initially for the slope calculation because the line had been drawn to go through these points.

A more general example of determining the slope of a line is illustrated in Figure 8. Note the following features of this example.

a. The points are chosen as far apart as possible. By making Δx and Δy large, the effect of any uncertainties in reading the values from the graph is minimized.
b. The points used for the slope determination are clearly marked and the coordinates are written beside the points.
c. The slope calculation is presented, so the result can easily be checked by the experimenter or by someone else at a later time.

The ability to determine the slope of a straight line is important because often the value of a slope can be related to a fundamental property. For example, the variation in vapor pressure of a

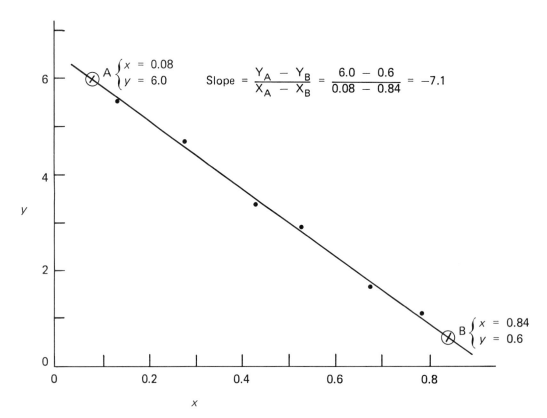

Figure 8. Determining the Slope of a Linear Graph

liquid as a function of the temperature is given by a well-known theoretical equation. One of the factors entering into this equation is the heat of vaporization of the liquid—the amount of energy needed to evaporate a mole of the liquid. It is possible to express the theoretical equation in the form of the equation of a straight line, and the slope is—(*heat of vaporization*)/2.303R, where *R* is a known constant. Thus, we can plot the experimental vapor pressure-temperature data appropriately, measure the slope of the line, and finally calculate the heat of vaporization from

$$heat\ of\ vaporization\ =\ -2.303R\ (slope)$$

This particular case is considered further in Experiment 4 of Chemistry 14, and other cases of utilizing the slope of a straight line occur in Chemistry 15.

Graphs with Curved Lines

Some functional relationships which are important in chemistry do not agree with the equation for a straight line. In many cases, however, a mathematical transformation is possible that converts the non-linear relation into a linear function. Chemists often look for such alternative ways to express relationships, since graphs of straight-line functions are easier to plot and to interpret.

Let us consider an example of the simplification possible by converting a non-linear function into a linear one. The familiar relation between the pressure and volume of a sample of gas at a constant temperature—Boyle's Law—is expressed mathematically by the equation

$$PV\ =\ c$$

where c is a constant whose value is determined by the temperature and the size of the sample. This equation matches that of a hyperbola, and thus give a curved line. A plot of some typical data is shown in Figure 9(a), where V is plotted as a function of P. It obviously would be difficult to extrapolate using such a graph.

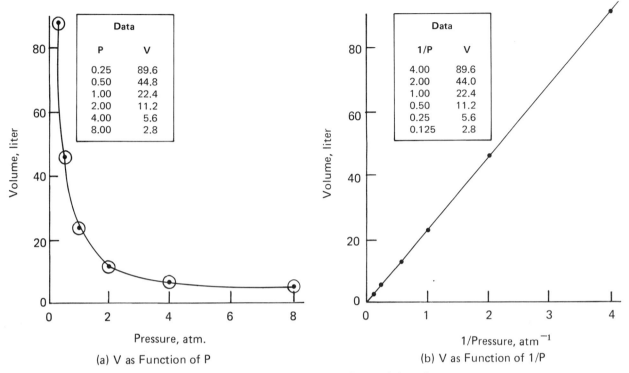

Figure 9. Alternative Plotting of Boyle's Law Data.

Note, however, that the Boyle's Law equation may be easily rearranged to give

$$V = c \left(\frac{1}{P} \right)$$

$$y = m x + b$$

This equation now is of the form of a straight line, where the x-variable is $(1/P)$. If the same data are replotted with V as a function of $(1/P)$, the graph of Figure 9(b) is obtained. Note that the line goes through the origin, since the equation requires that $V = 0$ when $1/P$ is zero. This linear graph is certainly easier to use for both extrapolation and interpolation.

As another illustration of this approach, consider the more complicated equation relating the equilibrium constant (K) to the absolute temperature (T).

$$-RT \ln K = \Delta H - T \Delta S$$

Clearly, if K were plotted as a function of T, a non-linear graph would result. [ΔH and ΔS are to be considered as constants.] But note what can be done if we rearrange the above equation, first by dividing through by RT

$$-\ln K = \frac{\Delta H}{RT} - \frac{T\Delta S}{RT}$$

$$\ln K = -\frac{\Delta H}{R}\left(\frac{1}{T}\right) + \frac{\Delta S}{R}$$

This expression now has the form of the equation of a straight line. Plotting $(\ln K)$ vs. $(1/T)$ will yield a linear plot with slope of $(-\Delta H/R)$ and a y-intercept of $(\Delta S/R)$.

Not all non-linear relations need be transformed into linear functions, and indeed many cannot be so converted. In these cases graphs with curved lines can be used directly to obtain useful information. Examples of the use of curved-line graphs occur in Experiment 1 of Chemistry 14 and in several experiments in Chemistry 15.

Laboratory Glassware for Volume Measurements

The measuring and dispensing of known volumes of liquids occur extensively in chemical experimentation. A variety of specialized containers are used for the measurement, depending upon the type of experiment and the precision which is required. Some of the kinds of volumetric glassware are illustrated in Figure 10.

Graduated cylinders are normally used when high precision in volumes of reagents is not needed, for example in many synthetic experiments. Cylinders, however, are not calibrated in the analytical sense and should not be used in place of pipets or volumetric flasks for the preparation of solutions of precisely known concentrations. A cylinder can normally be read to ± 1-2% of the total capacity of the cylinder. Glass graduated cylinders are commonly fitted with a plastic "bumper" to protect against breakage if the cylinder is knocked over on the benchtop. To be effective, the bumper guard must be placed near the top of the cylinder. The guard is easily removed when the cylinder is washed.

Burets are designed to facilitate the delivery of a precisely known amount of liquid. A buret consists of a long, small-diameter tube graduated with volume markings, and fitted with a stop-cock at the lower end for controlling the flow of liquid from the tube. Common sizes are 25 ml and 50 ml, with scales sub-divided into 0.1-ml divisions. Volumes can thus be estimated to one or two hundredths of a milliliter. The 25-ml buret commonly used in our course allows a volume of 10 ml to be measured with better than 0.5% precision. Larger volumes can be measured even more precisely.

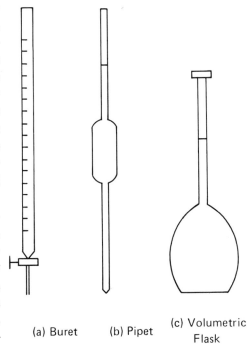

(a) Buret   (b) Pipet   (c) Volumetric Flask

Figure 10. Volumetric Glassware.

Pipets of the volumetric or transfer type are used for the delivery of a fixed, precisely measured volume of liquid. Commonly used sizes range from 1 ml to 100 ml. The volume is indicated by a single etched line around the upper stem of the pipet. Careful use of a volumetric pipet allows a reproducibility of the delivered volume of at least one part per thousand (0.1%).

Volumetric flasks are designed to facilitate the preparation of a solution of precisely known volume. The capacity is marked by a single etched line around the elongated neck of the flask. The narrow diameter of the neck means that any slight error in adjusting the total liquid volume to the calibration line will result in a relatively small error. Volumetric flasks are commonly used in capacities ranging from 10 ml to one liter, with both smaller and larger sizes also available. Volumetric flasks also allow reproducibility to at least 0.1% (1 ppt.).

Beakers, widely used as laboratory vessels, are not considered to be volumetric glassware. They are, however, often graduated with volume markings that are useful if a rough volume estimate is all that is needed. Volumes can be estimated to about ± 3-5% of the capacity of the beaker; for example, when using a 250-ml beaker, you could estimate the volume to ± 8-12 ml. Since this limit would apply to any volume you measured with the beaker, the percentage error would become large if you measured a small volume with this beaker.

General Use of Volumetric Glassware

Since volumetric glassware such as burets and pipets are used for measurements of high precision, it is essential that they be carefully cleaned. Otherwise the liquid will not drain uniformly, but will form irregular droplets in a non-reproducible fashion, thus limiting the calibration accuracy. The test of properly cleaned glass is that water will drain smoothly leaving only a uniform, nearly invisible film without droplets. The details of achieving this sort of cleanliness will be presented in the next section.

The water film adhering to clean glass has two important consequences for use of the glassware. First, not all of the liquid in a container can be poured out. The adhering film will drain slowly, and some liquid remains on the glass. This phenomenon means that the volume contained and the volume delivered are different. Consequently, volumetric glassware must be marked either TC for "to contain" or TD for "to deliver." A vessel marked TC is calibrated so that when it is filled to a graduation line it contains the specified volume. Volumetric flasks are manufactured on this basis. An object graduated and marked TD will contain slightly more than the marked volume, but it is designed to deliver the proper specified volume after a brief period of draining. Burets, pipets, and cylinders are generally calibrated to deliver. Details of usage reflecting the differences in calibration will be discussed in the sections on the different pieces of glassware.

The second result of the wetting of clean glass by water is that the flat surface of a liquid at rest is distorted where the liquid contacts its container. This distortion is especially pronounced when the container is a narrow tube like a buret or cylinder. The liquid-air interface assumes a definite curvature and is called a **meniscus**. For water and most other liquids in glass, the curvature is concave upward. The liquid level used to read the volume is taken as the *bottom* of the meniscus, since this level can be located more precisely. Liquids like mercury which do not wet glass give a meniscus which is convex upward, and in these cases the top of the meniscus is read.

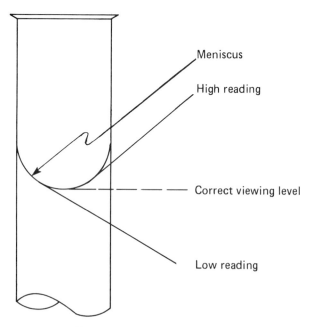

Meniscus

High reading

Correct viewing level

Low reading

Figure 11. Viewing a Meniscus.

When making a volume reading, the experimenter must take care to have his eye on a level with the meniscus bottom. If his eye is below the liquid level, the reading observed will be too small. If the eye is above the meniscus, the reading will be too large. These effects are shown in Figure 11. The variation of a reading with the position of observation is a general phenomenon called **parallax**. In this case it results from the fact that the scale (on the outside of the vessel) and the point being measured (in the center of the bore of the vessel) are not in the same plane.

Cleaning Volumetric Glassware

All glassware for volumetric measurement must be carefully cleaned if it is to yield data of the precision possible with the apparatus. Even traces of grease or other contaminants will cause a non-uniform drainage and the formation of water droplets which destroy the calibration. A careful worker will normally empty, clean, and rinse each item immediately after it is used and always before it is stored.

Cleaning is accomplished by rinsing out any contents, using a detergent, rinsing with tap water, and finally rinsing with distilled water. The result should be a surface which is covered uniformly by a thin, almost invisible film of water. If you can see discrete droplets of water on the inside surface, the cleaning is incomplete and must be repeated.

Use a warm dilute (2%) solution of detergent to cleanse the glassware. Be sure the solution contacts all the interior surface of the vessel. If a brush is used, be careful that exposed wire does not scratch the glass. Volumetric flasks can be cleaned merely by adding a small amount of the warm detergent solution and shaking the stoppered flask. To clean a buret, partially fill it with the detergent solution and allow some to run out through the tip. While tilting the buret toward the horizontal, rotate it about the axis to wet the entire interior. If this does not remove all grease, use a long-handled buret brush to scrub out the bore. For cleaning a pipet, draw in the detergent solution (use rubber bulb) until the pipet is about half full. Then hold the pipet horizontal and rotate it so that all surfaces are wet by the detergent solution. Be sure the solution enters the upper stem above the graduation mark.

The detergent washing must be followed by a thorough rinsing. First, rinse several times with tap water. The most effective rinsing is done using successive small portions rather than one large portion. Finally, rinse the glassware with distilled water. Again, use three or four small portions; for example, to rinse a 100-ml volumetric flask, add about 5-10 ml distilled water, rotate and tip the flask to cover all interior surfaces, and then pour out the liquid. Repeat with two additional portions. Never fill a container completely with distilled water to rinse it, since that would be wasteful of a valuable material.

Volumetric glassware will not need to be dried before use in your work, since it will be used only with aqueous solutions. All that will be necessary will be to rinse carefully with the solution being introduced so that concentrations are not changed. If non-aqueous liquids were to be used, the glassware would be dried by rinsing with acetone and air drying. Never attempt to dry volumetric glassware by heating with an open flame.

Using a Pipet

Liquid is drawn up into a pipet by means of a rubber suction bulb placed on top of the stem. *Never* use your mouth to provide suction to fill a pipet.

First, clean the pipet as outlined in the section on cleaning glassware. The tap water rinse can be accomplished by letting tap water flow into the stem and through the bulb, provided you periodically allow the bulb to drain empty. Be sure all detergent is washed out. Then rinse the pipet with three small portions of distilled water. With each, tilt and rotate the pipet so all interior surfaces are washed.

Now the pipet must be rinsed with the solution to be measured. Draw some of the solution into the bulb and thoroughly wet the interior surfaces as you did in the cleaning process. Be sure to rinse the stem to above the calibration mark. Discard this solution, and repeat the process with two more portions of the solution. You may now assume that all of the water originally present has been displaced by the solution and you are ready to measure a portion for transfer.

If you are right-handed, grasp the stem of the pipet in that hand between the thumb and last three fingers so that the tip of your index finger can reach the top of the stem. Partly compress the rubber bulb in your left hand and hold it against the top of the stem. Then with the tip of the pipet well below the surface of the liquid, gradually release the bulb to draw the liquid up to above the calibration mark. Remove the bulb and quickly place your index finger on the stem opening to trap the liquid. The liquid level should still be above the calibration mark. If it has dropped below, the bulb must be used again. When using the bulb, be careful not to draw the liquid into it.

With the pipet filled and your finger still on the top, remove the liquid from outside the tip by wiping with a clean tissue or by rinsing it off with a stream from your wash bottle. (The tip need not be dry but all excess solution must be removed.) Now touch the tip to the inside of a clean beaker and allow the liquid to drain slowly until the bottom of the meniscus coincides with the etched graduation line. Slightly rotating the pipet against the finger as it is relaxed may help in controlling the flow. Touch the tip to the inside of the beaker to remove any hanging drop. Now, carefully move to the new vessel to receive the sample, hold the tip against the inside of the vessel, and allow the liquid to drain.

After the flow has apparently stopped, wait 20 seconds with the pipet vertical to provide for reproducible draining. Touch off any adhering drop, but do *not* blow out the small amount of liquid still inside the tip. The magnitude of this small quantity is reproducible and was allowed for when the pipet was calibrated.

Using a Volumetric Flask

Volumetric flasks allow the preparation of solutions of known concentration since the total volume is known precisely. A solid solute can be weighed out and then transferred quantitatively to the flask before dilution to the mark. In other cases the solute is added in the form of a quantity of a more concentrated solution which is delivered into the flask from a buret or a pipet. In either case, the correctness of the final solution concentration depends upon the proper dilution to the mark on the flask.

First, the flask must be cleaned and rinsed as described previously. After the solute is transferred, the solvent is added until the flask is about half full. It is then agitated to mix the contents, and in the case of solid solute to dissolve the solid. The flask should be held by the top of the neck so as not to warm and expand the liquid by the heat of your hands.

When the contents appear homogeneous, more solvent is added until the bulb of the flask is almost full. The contents should be agitated by swirling to give mixing before the final dilution. (There may be appreciable volume change on mixing, so that partial mixing should be accomplished before the flask is completely full.) Add solvent until the level is close to the calibration line, again with swirling to avoid later changes of volume on mixing. Allow time for liquid to drain in the neck and then make the final addition of solvent to bring the bottom of the meniscus to the calibration mark. Remember to view the mark at eye level to avoid parallax error. When the eye is at the proper level, the front and back lines of the encircling mark will merge. The final addition can best be made with a medicine dropper or a plastic squeeze bottle (wash bottle).

Finally, the last solvent added must be mixed thoroughly with the rest of the contents. Mixing in a filled volumetric flask is quite inefficient. The stoppered flask should be inverted, and re-righted, with shaking, a minimum of twenty times to ensure a uniform mixture.

Using a Buret—Titration

A buret is often used to perform a **titration**, in which a reactant called the **titrant** is added from the buret to a reaction vessel until the desired reaction is complete. The completion of the reaction, called the **end point**, is detected by a change such as a color change of some substance called an indicator in the reaction vessel. The technique of titration is described below.

First, the buret must be properly cleaned and rinsed. It is then rinsed with three small (3-4 ml) of the solution to be used in the buret. With each portion, open the stopcock to let a few drops of the solution flow out. Then, while holding the buret almost horizontal, rotate and tilt it so that all interior surfaces are wet by the titrant solution. Continue to rotate the buret about its axis while tipping it farther so that the rinsing solution runs out the top of the buret into the sink. After the three rinsings, you may assume the liquid remaining in the buret is undiluted and the buret may be filled.

Close the stopcock and fill the buret to above the top graduation mark. Open the stopcock wide to expel any air bubbles in the tip. The tip must be completely filled with titrant solution both before and after a titration if the volume measured is to correspond to that delivered. The liquid level should be lowered to be below the zero mark, but do *not* waste time trying to set it *exactly* to 0.00 ml. Remove any pendant drop of liquid from the buret tip and record the initial buret volume reading.

In order to achieve the precision possible with a buret, it is necessary to read the meniscus carefully. The precise location of the bottom is made evident if a dark background is placed behind and just below the liquid level in the buret, as shown in Figure 12. Prepare a **meniscus illuminator** for this purpose by drawing a line on a card or piece of paper and thoroughly blackening a strip 1/2 in.

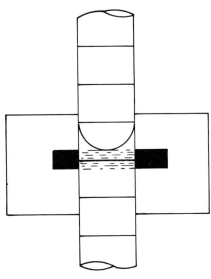

Figure 12. Using a Meniscus Illuminator.

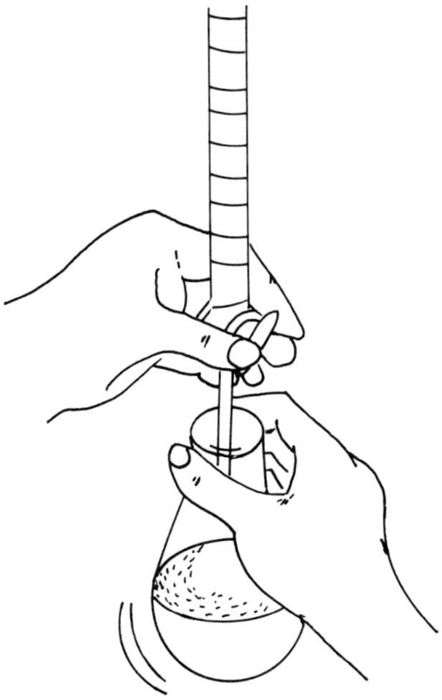

Figure 13. Titrating in an Erlenmeyer Flask.

below this line. To use the illuminator, hold it behind the buret and adjust it up and down until the bottom of the meniscus appears as a sharp black line. This will allow you to estimate the reading to the nearest 0.01 ml.

Read and record the initial buret volume. The usual vessel used for a titration is an **Erlenmeyer flask,** the sloping walls of which are designed to permit mixing without danger of splashing. The correct technique is suggested by Figure 13 for right-handed persons. Grasp and control the stopcock with your left hand. Hold the flask by the neck and make the liquid in the flask swirl. Do this by moving the bottom of the flask in a small circle while keeping the neck centered around the tip of the buret. The tip should extend inside the neck to avoid any loss of titrant solution. The buret should be clamped with the tip far enough above the bench top to provide clearance for easy handling and swirling of the flask. The flask should not set on the bench top during a titration.

The titrant is added until the end point is reached. In most cases the addition can be quite rapid until you see the indicator beginning to change color where the titrant enters the solution. Then the flow rate should be reduced, finally to a dropwise rate as the end point is reached. As the end point approaches, stop and rinse down the sides of the flask with a wash bottle stream to be sure all reactant is in the body of the liquid. A more precise location of an end point can sometimes be made by splitting drops of added solution. To do this, carefully open the stopcock to allow a small drop to form, much smaller than the ones which fall freely from the buret tip. Then touch the flask to the tip and rinse the droplet down the wall.

When the titration is complete, allow some time for final drainage in the buret to occur. When the level is stabilized, record the final volume. When you have finished all the titrations in a period, empty the buret, wash out the titrant, and rinse the buret with distilled water before storing it.

Use of the Analytical Balance

The analytical balances you will use are precision instruments costing over $1,500 each. They are designed to permit weighing to $\pm$ 0.0001 g, but to maintain this limit of precision they must be used with care.

Different laboratories will use different makes and models of balances, and thus the details of operation will vary. There are, however, a few common rules which apply to the use of any analytical balance.

A. Maintain a clean balance. Try to avoid all spills, but if you have an inadvertent spill (even a single crystal), IMMEDIATELY SWEEP OUT THE SPILLED MATERIAL.
B. Add or remove objects carefully—do not drop any object onto the balance pan.
C. Be sure the door is closed during a weighing to avoid the effects of air currents.
D. When you leave the balance, clean the pan, brush out the compartment, and close the balance doors.

You will have one of two main types of balances for your use. These differ fundamentally in construction and hence in operation. Brief descriptions are given below for each type.

Mechanical Suspension Balances

The heart of the balance is a suspension system which permits motion with as little friction as possible. It includes fragile knife-edges of synthetic sapphire which rest on polished plates of the same material. Any chipping or damage of the knife-edges will adversely affect the balance's performance.

The beam suspension is controlled by a three-position arrestment knob. One position locks the beam and removes the knife-edges from contact with the plates by raising the beam and the pan. A second, or partial release, position allows the beam to move somewhat but restricts the contact of the knife-edges. The third, full-release position allows the beam to move freely, resting on the knife-edges. All operation of the arrestment knob and the weight control knobs should be done slowly and smoothly to avoid damage to the knife-edges from impact when they come in contact with the plates.

Because different laboratories have different models of balances which vary in the locations of controls, read-outs, etc., you will receive detailed directions from your instructor on the use of the specific balances in your lab. The guidelines below, however, apply to all balances and should be reviewed before each use. They will serve to remind you of some of your instructor's comments and will help you follow good practice in the use of the balance.

1. Be sure the beam is arrested fully before adding an object to the balance pan or removing one.
2. Add or remove weights only when the beam is in either the arrested or the partial release position.
3. Make no changes in the loading of the balance when the beam is fully released. *Do not change weights or move an object on the pan when the arrestment knob is in the fully-released position.*
4. Lower the beam *slowly* when going from arrest to either release position.
5. Turn the weight knobs smoothly and slowly. Do not twirl them when changing the weights.
6. Turn all weight dials to zero when you are finished making a weighing.
7. Clean the pan, brush out the compartment, and close the balance doors before you leave.

Electronic Balances

No beam is used, but instead the balance pan is stiffly supported on a "strain cell". The detection device is an electrical circuit which responds to the very small motion of the balance pan when an object is added. The mass is then computed from the magnitude of an electric current that is proportional to the deflection in the cell. In addition to the general rules A–D above, specific details for use will be supplied by your instructor.

Use of the Triple Beam Balance

This balance is appropriate for weighing objects to about the nearest tenth of a gram. Triple beam balances are kept on the side shelves and should be used as follows.

1. With the pan empty and all three sliding weights at the zero position, check to see that the pointer hovers at the zero mark. If it does not, turn the adjusting screw (under the pan) until balance is achieved. If you are unable to achieve a proper zero reading, check with the instructor.

2. Place the object to be weighed on the pan. Do this fairly gently; the pan and beams are supported on a sharp pivot which can be blunted by misuse. Chemicals should not be placed directly on the pan. Liquids should be weighed in a pre-weighed beaker or flask. Solids may be weighed in a beaker or on a sheet of paper.

3. Move the heaviest weight (on the central beam) to the right from notch to notch until the pointer on the right end of the beam drops to the down position. Then return this weight one notch to the left so the pointer goes up again. Now do the same with the weight on the back beam. Finally, slide the smallest weight to the right until the pointer moves down and oscillates up and down around the zero point. These oscillations should be "symmetric" around zero when the sliding weight is properly positioned. (You need not wait for the oscillations to stop.) The weight of the object is the sum of the weights indicated on the three balance beams.

4. Record the weight to the nearest tenth of a gram directly into your laboratory notebook. (You could estimate the weight to the nearest hundredth of a gram, but the balance is too inaccurate for the last figure to be meaningful.)

5. Remove your object from the pan gently, but do not return all the weights to zero. (It is better not to leave the balance at zero, where it can teeter back and forth on its knife-edges.)

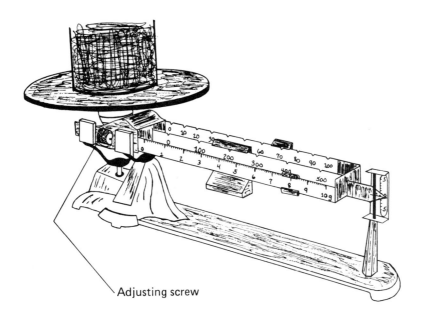

Adjusting screw

Figure 14. Triple Beam Balance Reading 278 Grams.

39

Experimental Section

TIME SCHEDULE

A suggested laboratory schedule involving 5 topics and 15 lab meetings is given below, but this may be modified by your instructor.

The time allocation for the five topics and 11 laboratory meetings of 4 hours each is given below. If your course is given in 15 3-hour meetings, a modification of the time allocation will be made.

In either event, the order will be determined by your instructor.

Topic	Number of Lab. Meetings
Check-in; Laboratory Techniques	1
Experiment 1	1
Experiment 2	2
Experiment 3, Parts I and IIA	1
Experiment 3, Parts IIB-C	1
Experiment 3, Completion	1
Experiment 4	2
Experiment 5 and Check-out	2

Check-in and Laboratory Techniques

In this topic and in others, you will have to use your time efficiently in order to complete your work. For example, in this topic do not waste time waiting in line to use a balance; instead, work on the glass bending until a balance is free for use.

After the check-in procedure is completed, you will proceed with exercises to acquaint you with the use of the analytical balance, the triple-beam balance, and the laboratory burner, and to introduce you to some techniques of working with glass. As a part of this, your instructor will demonstrate for you:

a. How to use the analytical (Mettler) balance. (Demonstrated in small groups.)
b. How to handle flasks and test tubes with a paper collar to avoid affecting their weights.
c. Proper adjustment of the Bunsen burner. (If your own burner proves defective, consult your instructor about having it replaced.)
d. Proper technique for cutting glass tubing. (Be sure your own glass scorer is sharp. If it is not, obtain a new one from the stockroom before you attempt to cut your own tubing.)
e. Methods for polishing, bending, sealing off, and flaring glass tubing, and for drawing capillary tips.
f. The safest method for inserting glass tubing into rubber stoppers.

In these first laboratory periods you will make the items shown in Figure Tc-3 on page 46 to be used in later experiments. The glassworking techniques you will need are described after the followed discussion of the Bunsen burner.

Use of the Bunsen Burner*

Observe that your burner has a knurled knob at the very bottom. This **needle valve** controls the amount of gas entering the burner from the gas line. The barrel of your burner is screwed onto the base. By twisting the barrel, you can cause the holes near the bottom of the barrel to become more open or more closed. These are air ports, and twisting the barrel enables you to adjust the amount of air that enters the barrel to mix with the gas. Adjusting a burner to give a hot flame always involves adjusting gas and air controls to achieve an optimum mixture.

Unscrew the barrel completely from the burner and observe the small orfice through which the gas enters from the base. To demonstrate the nature of the gas flow, light the gas while the barrel is off according to the following instructions. Turn the needle valve clockwise until it stops (fully shut) and then open it about one-quarter turn. Attach the rubber tubing to one of the gas outlets on your desk. Have a match handy. Now turn the gas cock full on by moving the handle until it points parallel to the nozzle. (If you move it beyond that point, it begins turning the gas off again.) Strike your match and light the gas coming from the small orfice. It may take a few seconds to sweep all the air out of the rubber tubing before the gas will ignite, and you may have to reduce the gas flow to get a stable flame. Since the gas has not been pre-mixed with air, the flame is yellow and relatively cool. Sometimes a flame of this sort is used for warm small objects.

Turn off the gas, reassemble the burner, and prepare to ignite it to get a hot flame. First open the needle valve about one turn from its fully-closed position. Adjust the barrel until the air ports are about one quarter open. Now, with a match handy, turn the gas cock on the desk to full on. Strike the

---

*There are several varieties of laboratory burner, all of them loosely referred to as Bunsen burners. Strictly speaking, those in use at Penn State are *Tirrill* burners.

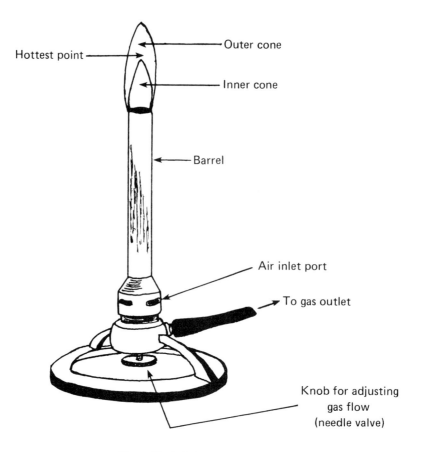

Hottest point

Outer cone

Inner cone

Barrel

Air inlet port

To gas outlet

Knob for adjusting
gas flow
(needle valve)

Figure Tc-1. Laboratory Burner.

match and bring it slowly *from the side* to a position just at the top edge of the barrel. (If you thrust the match directly over the burner, the stream of gas may merely blow the match out.)

If the gas tries to ignite but does not sustain a stable flame, it is likely that the amount of air being admitted is too *great* for the amount of gas entering; shut off the gas cock and either increase the gas supply by turning the knurled knob more turns counter clockwise, or else reduce the amount of air by screwing the barrel down, or both. If you are still unable to get a stable flame, consult your instructor.

If you do get a stable flame, examine it to see if it is yellow and flickery, like a candle flame. If it is, then too *little* air is entering the barrel. Without shutting off the flame, twist the barrel of the burner to admit more air. As you do this, the flame should turn blue, stand erect, and produce a faint rushing or roaring sound. Too much air results in a flame which is less stable. Experiment with the air adjustment until your flame is satisfactory. If you decide your flame is too large or too small, adjust the gas supply (knurled knob) to suit. This will require that the air be readjusted also. You will need to readjust the burner during your experimental work to match the flame to the requirements of the job at hand.

Sometimes an excess of air will cause the gas mixture to burn so rapidly that the flame travels down the burner barrel (**flashes back**) and burns at the base of the burner. This is a dangerous situation because (1) the barrel becomes extremely hot and (2) the base will eventually become hot enough to soften the rubber tubing, leading then to a fire at that point as gas escapes. You should be alert to a "flash-back" and immediately turn off the gas. Reduce the amount of air entering the burner before you attempt to relight it.

When a burner is properly adjusted, the flame will be seen to consist of two parts—an inner bright blue cone surrounded by a much paler outer cone. The inner cone consists of *cool* unburned gases, while the outer cone is the zone of burning. The hottest part of the flame is at the point just *above* the tip of the blue inner cone. You can demonstrate the relative temperature of the two cones by the following test. Adjust the burner to give a flame about 1 1/2 inches tall with a well-defined inner cone. Light a match and immediately extinguish it so that you have the unburned wood. Hold the match horizontal (use forceps to avoid burning your fingers) and place the match so that it passes through the flame about 1/2 inch above the top of the barrel. In 1-2 seconds remove the match, blow it out if it is afire, and observe where it is charred.

Glass Working

Cutting Glass Rod or Tubing. After measuring to determine the location of the cut, make a decided scratch in the tubing using a glass scorer or a three-cornered file. This should be done in one or two firm thrusts, not in a sawing motion. (If your scorer is not sharp, obtain a new one from the stockroom.) Holding the glass in a towel to protect your fingers from jagged edges and with the scratch *away* from you, place one hand on either side of the scratch. Your thumbs should press against the glass on the side opposite the scratch. Now bend the ends of the glass back toward you using your thumbs as a fulcrum. The glass should break cleanly at the scratch.

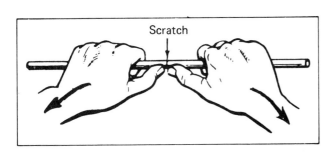

Figure Tc-2. Breaking Glass Tubing.

Fire Polishing Glass Tubing. The ends of freshly cut glass are sharp and both difficult and dangerous to insert into rubber stoppers or tubing. They should be rounded smooth by fire polishing. Adjust your burner to obtain a stable blue inner cone of flame. Slowly lower the end of the glass to be polished into the flame. (Overly rapid heating, or cooling, may crack the glass.) Rotate the glass with the end just over the apex of the inner blue flame. As the glass begins to melt, an orange-yellow plume of flame, characteristic of sodium ions in the glass, will occur. Continue to rotate the glass in the flame until the sharp edge of the glass itself has a *very faint* orange incandescense. Remove the glass, and examine it. The end should appear smooth and slightly rounded. Place the hot end on an asbestos pad or wire gauze so it can cool slowly. **Do not hand hot glass to your instructor for inspection.**

Glass remains hot for a long time. To test glass for coolness, pick it up by the cold end and move a finger *slowly* along the glass towards the possibly-hot part. In this way, you can ascertain if the glass is still hot without buring yourself.

Bending Glass. If glass tubing is heated in a normal burner flame and then bent, the walls will be pinched in at the bend. To achieve a smooth bend in which the original diameter of the bore remains constant, it is necessary to heat tubing uniformly over an extended length of 1 1/2-2 inches. A "wing-top" **flame spreader** (sometimes called a "fish-tail" spreader) provides the broad, flat flame.

Attach a wingtop flame spreader to the top of your unlit burner. Turn on and adjust the burner to give a stable blue flame. Heat the section to be bent by lowering the glass slowly into the flame until it is at the top of the blue flame. Rotate the glass slowly with your fingers and also move it back and forth horizontally in the flame to extend the length being heated. As the glass softens, you will need to hold it carefully to keep it fairly straight. When it feels fairly flexible, *remove* the glass from the flame and bend it to the desired angle. Do this quickly since the glass becomes rigid after a few seconds. Hold it motionless until it has become firm. Allow the bend to cool on an asbestos pad or wire screen.

Making a Capillary Tip. Rotate a section of glass over the hot part of the burner flame (*without wingtop*) until the heated portion feels very soft and flexible. Remove it from the flame and draw your hands apart smoothly. You should obtain a capillary section joining the two ends of tubing. *After cooling*, nick the capillary with your scorer and *very gently* break the glass (it will almost fall apart). Do not fire polish the capillary tip.

Flaring the Ends of Glass Tubing. To make the end of your tubing flare out so that it will firmly hold a rubber bulb, heat and rotate the end in the burner as though you were fire polishing it. Keep it in the flame until it is quite soft, and then press it firmly against your asbestos pad. The end will bulge. Repeat this procedure until the bulge is sufficiently large.

Sealing the Ends of Glass Tubing. Heat a mid-section of tubing as though you were going to make a capillary. When it is soft, keep it in the flame and pull away one end of the tubing. (Keep the end to be sealed in the flame.) The glass should separate, leaving a pointed end on the half in the flame. Continue heating and rotating this end in the flame until it has become somewhat rounded and thickened. If a small glob of glass remains on the end, ignore it.

Inserting Glass into Rubber Stoppers and Tubing. Glass to be inserted should **always** be fire polished. To lubricate the joint, place a drop of glycerine (side shelf) on the glass and spread it around with a fingertip. Grip the glass through a protective towel, holding the glass *near* the end to be inserted. Hold the stopper or tubing in your other hand and push the glass into the hole with a twisting motion. DO NOT FORCE! A serious cut due to glass breaking while being inserted into a rubber stopper is the most common student injury.

Making Glass Items. The following items, needed for future work, should be made from the glass tubing provided at check-in.

1. Three right angle bends. Two of these should have equal arm lengths of 2 inches. (For use in Experiment 2.) The third should have unequal arm lengths of 2 inches and 5 inches. (For use in the trap bottle of Experiment 4.) (See figure.)
2. One capillary eye dropper. Flare the other end so that it fits the rubber bulb on one of the regular eye droppers in your desk (see figure). (For use in Experiment 4.)

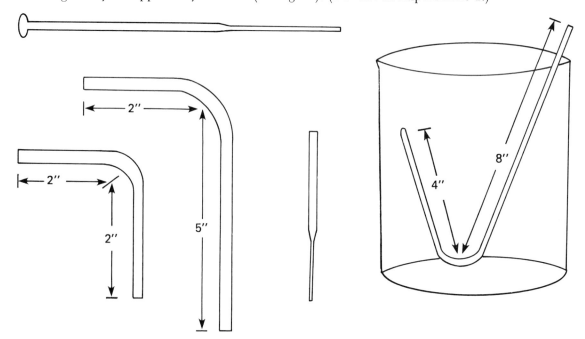

Figure Tc-3. Glass Items to be Made.

3. One capillary tip. Use the other half of the capillary you drew for the eye dropper. Do not flare the other end of this piece. Cut it so that the heavy glass stem is about 2 inches long. Fire polish the cut end. (For use in Topic 4.) (See figure.)

4. One vapor pressure tube (isoteniscope). *First*, make a sealed end and, when it has cooled, make a 40° bend about 4 inches back from the seal. Cut the glass so that the open end is 8 inches long. Fire polish the open end. The isoteniscope should fit in your largest beaker as sketched in the figure. (For use in Topic 4.)

Testing the Isoteniscope. When the isoteniscope is cool, you should test it to see if the end is sealed. The instructor will have in the laboratory a piece of heavy walled rubber tubing (vacuum tubing) attached to an aspirator (a suction device which is attached to a water faucet). Put the open end of your isoteniscope into the vacuum tubing (use glycerine), and submerge the closed end in a beaker of water. Turn on the aspirator by opening the faucet. If your isoteniscope is not properly sealed the aspirator will cause water to leak into the "closed" end. If your isoteniscope leaks, you must dry it before reheating it or it will crack. Pour out any water, and use the aspirator to draw air through it until all moisture is gone. When reheating, lower the end very slowly and with rotation into the flame.

Weighing on the Analytical Balance

As you do the following test, you should record the observed data *directly* in your laboratory notebook.

You should test your ability to use these balances by

a. weighing your small (clean, dry) test tube;

b. weighing a cork (any size);

c. weighing the small test tube and cork together;

d. checking that the sum of the weights of test tube and cork is equal to (within 0.5 mg) the weight of test tube and cork together;

e. estimating the uncertainty in a weighing by repeatedly (3-5 times) reweighing a given object. A weighing bottle is an appropriate item. Handle it only by a paper band so as not to transfer materials from your fingers. Likewise set the bottle down only on a clean piece of paper so it does not pick up dirt from the balance shelf. Return all weight dials to zero and check the zero position of balance before each weighing.

Enthalpy of Reactions: Hess's Law

The purpose of this experiment is to determine heats of neutralization and solution using a simple calorimeter. Combining these with other data as indicated by Hess's law will allow you to compute the heat of formation of a chemical compound.

Theory

The **enthalpy** of formation, ΔH, of a chemical compound is defined as the energy absorbed when the elements react to form the compound.

Let us consider the heat of formation of HCl. If we measure the energy released in the (exoergic) reaction of one mole each of H_2 and Cl_2 (both initially at 1 atm. and 25°C) to produce two moles of HCl at the same conditions, we would find 184.62 kilojoules (kJ) of energy. This is 92.310 kJ per mole of HCl produced, so the heat of formation of HCl, symbolized ΔH_{298}, is −92.310 kJ/mole. **Exothermic** processes are associated with negative enthalpy changes.

Hess's law states that, if a process can be broken down into a series of steps, ΔH *for the over-all process will be the algebraic sum of the ΔH's for the individual steps.* For example, suppose we want the heat of formation of CO and we know the heat of formation of CO_2 as well as the heat of combustion of CO:

$$C(s) + O_2(g) \rightarrow CO_2(g) \qquad \Delta H = -394 \text{ kJ}$$
$$CO(g) + 1/2\, O_2(g) \rightarrow CO_2(g) \qquad \Delta H = -284 \text{ kJ}$$

We reverse the second reaction and add it to the first:

$$C(s) + O_2(g) \rightarrow CO_2(g) \qquad \Delta H = -394 \text{ kJ}$$
$$\underline{CO_2(g) \rightarrow CO(g) + 1/2\, O_2(g) \qquad \Delta H = +284 \text{ kJ}}$$
$$C(s) + 1/2\, O_2(g) \rightarrow CO(g) \qquad \Delta H = -110 \text{ kJ}$$

Thus, by constructing a two-step process which takes us from $C + O_2$ to CO, we find the heat of formation of CO to be −110 kJ/mole. We can represent the idea schematically in Figure 1-1.

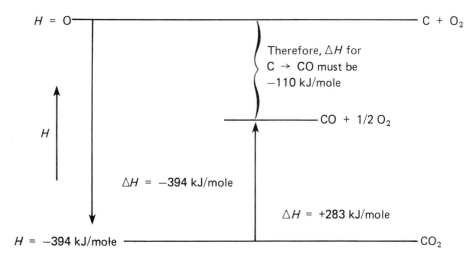

Figure 1-1. Enthalpy Diagram for Formation of CO.

In the first part of this experiment, you will determine the heat of formation of solid NH_4Cl. You will do this indirectly by measuring the heat of neutralization for the reaction

$$NH_3(aq.) + HCl(aq.) \rightarrow NH_4Cl(aq.)$$

and also the heat of solution in the process

$$NH_4Cl(s) \rightarrow NH_4Cl(aq.)$$

Combining these data with other enthalpy changes which will be provided, you will compute the heat of formation of solid NH_4Cl.

You will use a simple calorimeter to measure the heat evolved or absorbed in a given process. The **calorimeter** consists of an insulated reaction chamber into which dips an accurate thermometer. The reaction chamber contains a known mass of water and also the substances involved in the process being studied (dissolving, neutralization, etc.). If the process is exothermic, the released heat causes the water to become warmer. This is registered on the thermometer. Knowing the mass of water present and the increase in temperature, you can calculate how many joules of heat were released by the process occurring in the calorimeter. An essential relationship is that 4.184 joules of heat increases the temperature of one gram of water by one degree Celsius.

Measurement of the temperature change, ΔT, will require that you record your thermometer readings over a period of about 11 minutes and then plot the data as indicated below. The figure indicates that one measures temperatures for about 4 minutes before mixing in the reagent(s) and continues reading temperatures for about 7 minutes thereafter. Since there is some finite time involved in the completion of the process, the uniform heating of the water, and the registering of the temperature change by the thermometer, the curve has a shoulder which makes the temperature change hard to estimate. This difficulty is circumvented by extrapolating both the initial and final slopes back to the instant of mixing and allowing these intercepts to define ΔT.

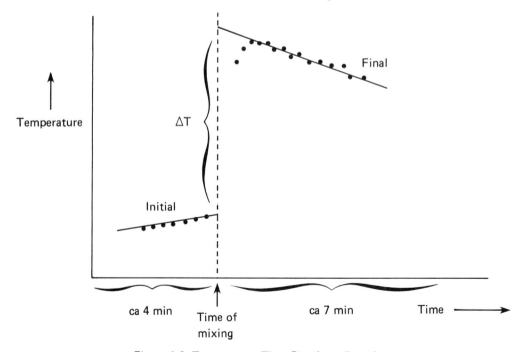

Figure 1-2. Temperature-Time Plot for a Reaction.

Experimental Procedure

Two students are to work together on this experiment. However, you will have individual unknowns and submit separate reports.

Remember—all data must be recorded directly in your lab notebook as they are observed.

You will be provided with four styrofoam cups and a thermometer which is calibrated to tenths of degrees Celsius. **This thermometer is quite expensive and should be handled carefully.** Cut the rim off of one cup so that, inverted, it fits into two nested cups as indicated in Figure 1-3. Make a hole in this same cup so that your thermometer can poke through. [One way to do this is by warming a glass rod in your burner and gently touching it to the cup.] Assemble your apparatus as shown, being careful not to cover over the thermometer scale from 20°-40°, which you will need to see to make your measurements. You will open and close your calorimeter by simply sliding the top cup up and down the thermometer.

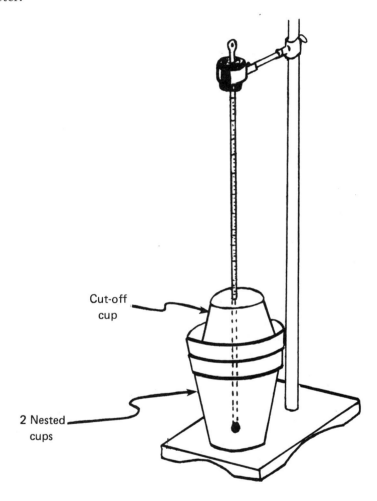

Cut-off cup

2 Nested cups

Figure 1-3. Calorimeter.

Heat of Neutralization. You will measure the temperature of the solution inside the calorimeter as a function of time. Your data must be recorded in neat tabular form as shown below.

TABLE 1
Heat of neutralization of 1.5 M NH_3 with 1.5 M HCl—
Variation of temperature with time.

	Run 1	Run 2
HCl initial temperature	—	—
Time (minutes)	Temp. °C	Temp. °C
0	—	—
0.5	—	—
1.0	—	—
—	—	—
*—	—	—
—	—	—
10.5	—	—
11.0	—	—

*Time of mixing.

Use a graduated cylinder to obtain 50 ml of 1.50 M NH_3 solution from the side shelf. Place this solution into your calorimeter. Also, place 50 ml of 1.50 M HCl (aq) in your spare styrofoam cup. Measure and record the temperature of this HCl solution before proceeding. Making sure your calorimeter thermometer bulb is immersed in the NH_3 solution, close the calorimeter by lowering the top cup to fit snugly into the bottom cup. Measure the temperature as precisely as possible at 30 second intervals for about 3 1/2 minutes, swirling occasionally. It is recommended that one partner read the clock and record the data while the other partner reads out the temperature. After taking your 8th temperature reading (corresponding to time t = 3 1/2 minutes) open the calorimeter, pour in all 50 ml of HCl, swirl the mixture thoroughly for about 40 seconds, then quickly close the calorimeter, and measure the temperature at time t = 4 1/2 minutes. Resume taking temperature readings (with occasional swirling) at 1/2 minute intervals for 6 1/2 additional minutes. Your run should take about 11 minutes: 3 1/2 minutes before mixing, 1 minute for mixing and swirling thoroughly and 6 1/2 minutes after mixing. Repeat the entire process with a second pair of 50-ml samples. For this second run, reverse the roles of the partners.

Heat of Solution. In your preceding runs, you ended up with 100 ml of 0.75 M NH_4Cl. Calculate how many grams of solid NH_4Cl this solution contained; then weigh out this much NH_4Cl on the triple beam balance. Place 100 ml of distilled water in your calorimeter, measure the temperature for 3 1/2 minutes as before, quickly dump in your solid NH_4Cl sample, and stir the mixture with a stirring rod rapidly for about 20 seconds until the NH_4Cl has dissolved completely. If possible, take the temperature at t = 4 minutes, and continue taking readings (with occasional swirling) at 1/2 minute intervals for 6 1/2 additional minutes. Repeat with a second run. Your data for the heat of solution must be presented in a table with the title:

Table 2. Heat of solution of NH_4Cl—Variation of temperature with time.

Heat of Solution of Unknown. The instructor will supply *each student* with a solid unknown. Weigh out 3-5 grams of your unknown and measure the heat of solution as above. Repeat with a second 3-5 grams. You should continue to work as a team when taking data on your individual unknowns.

Upon completing your measurements, clean and return your thermometer and styrofoam cups according to your instructor's directions.

Treatment of Data

First, refer to pages 25-28 describing the preparation of graphs. You will want to be able to determine as precise a value of ΔT as your data permits. Plot your data (T vs. time) as indicated in the THEORY section and estimate your ΔT values. [Note: If your HCl and NH_3 solutions were not initially at the same temperature, you will need to average them to obtain a correct initial temperature.] Assuming that your calorimeter contains 100 grams of water in each run, and ignoring all other absorbers of heat, these ΔT values are readily converted to joules released (or absorbed). For the NH_4Cl runs, you know how many moles of material were involved, so you can calculate ΔH *per mole* for the two processes:

a. NH_3 (aq., 1.5 M) + HCl(aq., 1.5 M) → NH_4Cl(aq., 0.75 M)

and

b. NH_4Cl(s) → NH_4Cl(aq., 0.75 M)

Given the following enthalpy data, use your ΔH values to calculate the heat of formation for NH_4Cl(s).

	ΔH_{298} in kJ/mole
$1/2\ N_2$ (g) + $3/2\ H_2$ (g) → NH_3 (g)	−45.77
$1/2\ H_2$ (g) + $1/2\ Cl_2$ (g) → HCl(g)	−92.31
NH_3 (g) → NH_3 (aq., 1.5 M)	−35.40
HCl(g) → HCl(aq., 1.5 M)	−72.84

The data on your unknown enable you to calculate the heat of solution in joules per *gram* of material.

Organization of the Report

Review pages 4-7 describing the Report.

Your report must be divided into sections. The first section, which is page 1, must include your name, partner's name, desk number and the experiment's number, title and purpose. It must also include a reference to the procedure, e.g., The procedure followed is described in "Chemistry 14 Experiment Write-ups, Revised 1978."

The second section must be titled "DATA." It will consist of your original data sheets of your tabulated measurements, identification of your unknown, weights of samples, etc. These will be followed by the graphs used in the determination of ΔT. Recall that a good graph must have a title, have its coordinates properly labelled and be easy to read.

The third section must be titled "CALCULATIONS." Do your calculations as follows.

1. Calculate ΔH of neutralization of (NH_3 + HCl) and ΔH of solution of $NH_4 Cl(s)$, for the amounts of materials used in this experiment.

   Recall that

   a. heat evolved by reaction = heat used to raise temperature of system

   $$= \Delta T \text{ (degree)} \times C_p \text{ (joule degree}^{-1})$$

   where C_p = number of joules required to raise temperature of system by one degree (C_p is the heat capacity)

   b. heat evolved by a reaction (exothermic reaction) has a ΔH which is negative.

2. Calculate the values of ΔH for the same two processes (neutralization and solution) but on molar basis.

3. Using the values of ΔH determined in (2) and the enthalpy data given above, apply Hess' Law to calculate the heat of formation of $NH_4 Cl(s)$.

To establish the identity of your unknown calculate ΔH of solution per *gram* of your unknown. Your instructor will have posted a list of compounds with their heats of solution. Compare your ΔH with those listed and on the basis of this comparison, identify your unknown.

The last section must be titled "ERROR ANALYSIS, DISCUSSION." The required error analysis and discussion will be explained to you by your lab instructor. You should also read the section of "Error Analysis" in the INTRODUCTION and pay special attention to the "Example of Error Analysis," page 18. The experimental quantities you measured, and which must be considered in your analysis, are the temperature change and the volume of liquid (which affects C_p). When you calculate the ΔH/mole, you must also include the uncertainty in the measurement of number of moles. This depends upon the precision in the weighing of the solids or the measurement of the volumes of acid and base.

In this error section, include an answer to the following question. As the solution warms up, some heat must go into warming up the cup, the thermometer, and dissolved chemicals. If these, in the aggregate, have a heat capacity equivalent to 3 ml of water, what error does this produce in your results? Would this error be random or systematic error? Justify your answer.

Questions

Be sure you can answer these when you come to lab. There may be a quiz.

1. Given the following data, calculate the enthalpy of formation of $H_2 O_2$.

   $$H_2 \text{ (g)} + 1/2 \ O_2 \text{ (g)} \rightarrow H_2 O(\ell) \qquad\qquad \Delta H = -286 \text{ kJ}$$

   $$H_2 O_2 \text{ } (\ell) \rightarrow H_2 O(\ell) + 1/2 \ O_2 \text{ (g)} \qquad\qquad \Delta H = -99 \text{ kJ}$$

2. During an actual run with the calorimeter, there are two important criteria which must be satisfied by a *correctly positioned* thermometer. What are these two criteria?

3. If a temperature vs. time curve gives a ΔT of $10.0°C$, and you assume that all the heat produced has been absorbed by 100 grams of water, how many joules of heat have been produced?

4. If you read a temperature rise from your graph as $8.78°C$ and you estimate the absolute error in determining that rise to be $0.05°$, what is the relative error in ΔT?

Gas Volume-Weight Relations in Reactions

The purpose of this experiment is to determine the amount of nitrite in a sample by measuring the amount of nitrogen gas evolved in a chemical reaction.

Theory

When sodium nitrite ($NaNO_2$) is dissolved in aqueous solution it dissociates to form ions. Nitrous acid (HNO_2) is a relatively weak acid, and when the sodium nitrite solution is combined with a solution of the strong acid sulfamic acid (HSO_3NH_2) almost all NO_2^- is changed to the undissociated HNO_2 and the following reaction occurs:

$$HNO_2 + NH_2SO_3^- \rightarrow N_2(g)\uparrow + HSO_4^- + H_2O. \tag{1}$$

Notice that part of the nitrogen comes from the HNO_2 and part from the $NH_2SO_3^-$. If HNO_2 is present in excess, the reaction ceases when all the $NH_2SO_3^-$ is used up, so the amount of N_2 evolved is a measure of how much $NH_2SO_3^-$ we started with. This experiment works in the reverse way: $NH_2SO_3^-$ is present in excess, so each mole of $NaNO_2$ we start with should lead to the production of one mole of N_2 gas. In the first part of this experiment you will test this stoichiometric relationship. In the second part you will use the relationship to determine the percent $NaNO_2$ in an unknown mixture.

This experiment is designed so that you can collect the gas evolved in the reaction and measure its volume, temperature, and pressure. Assuming ideality, the number of moles of nitrogen formed may be computed from the perfect gas equation of state. This calculation will be dealt with more fully in the last section of this write-up.

The nitrogen content of biological materials, such as amino acids and proteins, is often determined by adding excess nitrous acid (HNO_2) and a strong acid to the materials and measuring the amount of nitrogen gas evolved. This analytical method is called the **Van Slyke determination**, after a famous biochemist. In such a determination it is necessary that the reaction be stoichiometric (it must go to completion and only one gas may be evolved), and the gas must be relatively insoluble in any liquid with which it comes in contact. The present experiment differs from the usual biochemical Van Slyke determination principally in that an inorganic compound (sulfamic acid) is used as a source of part of the nitrogen instead of a biological material.

Experimental Procedure

NOTE: Before starting the experiment be sure you have read the "SUGGESTIONS FOR ORGANIZING YOUR NOTEBOOK" on page 58. Planning ahead of time helps in producing a neat, orderly record.

Compute in your notebook the amount of $NaNO_2$ that would be required to produce 40 ml of dry N_2 gas at room temperature and pressure according to the equation for the reaction:

$$NaNO_2 + NH_3SO_3 \rightarrow N_2(g)\uparrow + NaHSO_4 + H_2O. \tag{2}$$

Use the ideal gas law, $PV = nRT$, with R = 82.05 ml atm/mole degree. You will use approximately this amount of $NaNO_2$ in your runs with the pure salt. The volume of 40 ml was chosen to use most of the 50-ml capacity of the buret and thus minimize the relative error in the volume measurement.

[If you have to wait to use a balance, proceed to set up the apparatus as described below. When your balance is free, weigh out your sample of $NaNO_2$ as described in the next paragraph.]

Dry* and weigh a small test tube. Obtain in this approximately the amount of pre-dried $NaNO_2$ you calculated above. Reweigh, as soon as possible, so as to obtain the sample weight to an accuracy of 0.1 mg. If you have obtained too much $NaNO_2$, it will be necessary to remove a portion of the sample and reweigh. Dissolve the nitrite sample by adding 2 ml of distilled water to the test tube. It is convenient to place this test tube containing the nitrite solution in a marked beaker to prevent it from overturning until you are ready to place it in the larger tube.

Students should work in pairs in setting up apparatus and running pure $NaNO_2$ samples. (Share your data, but enter it in both notebooks.)

Obtain a 50-ml Mohr buret with nipple and a leveling bulb from your instructor. The buret should be clean so that it drains without leaving water droplets adhering to the walls. If necessary, clean it according to the instructions on pages 34-35.

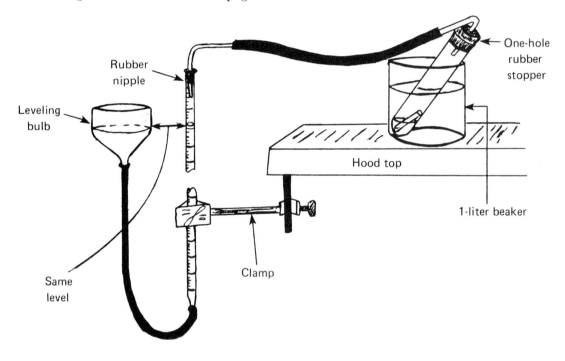

Figure 2-1. Apparatus for Collecting and Measuring Gas.

Prepare a large beaker full of water at room temperature (to within a degree). Be sure to record the room temperature in your notebook. Assemble the apparatus as sketched, but do not put the small test tube or any solution into the large test tube yet. Use glycerine when inserting

*Do not use compressed air from the desk line for drying glassware. The air from these lines usually contains moisture and oil from the compressor, and will deposit a dirty film on the glass. Instead, rinse the glassware with distilled water and warm it over a burner flame. Be sure to allow the glassware to cool before you use it.

glass into rubber stoppers or tubing. **Do not force glass rods into stoppers! Your instructor will demonstrate the proper technique for inserting glass into rubber.** All connections should be snug enough to be airtight. Be sure that neither piece of rubber tubing is pinched or folded shut. Open the system by removing the rubber stopper from the large test tube. Support the leveling bulb at a level near the top of the buret and pour room-temperature water into it until the buret is nearly full. Now insert the rubber stopper into the large test tube. By changing the height of the leveling bulb, you can change the total pressure within your apparatus. When the water levels in leveling bulb and buret are identical, the pressures are identical inside and outside the system. Be sure you understand the relation between pressure and the difference in water column heights.

To test for leaks, raise the leveling bulb to its maximum height, thereby increasing the pressure inside the apparatus. Observe the liquid level inside the buret for a minute or so. If no change occurs, the apparatus is airtight. If the level rises slowly, you must find and correct a leak. To locate the leak, pinch the rubber tubing which connects the test tube to the buret. If the level stops rising, your leak is in a connection near the test tube. If it continues rising, it is in a connection near the buret. Do not proceed with the experiment until your apparatus is leak-free. Then return the leveling bulb to a level near the top of the buret.

Remove the large test tube from the apparatus and place in it 0.5 g of sulfamic acid and 10 ml of distilled water. Swirl to dissolve. Then lower the small test tube containing the nitrate sample into the large test tube, being careful to avoid letting the nitrate and acid come in contact. Reconnect the large test tube to the apparatus and let it rest in the l-liter beaker as shown in the sketch. *Carefully* pour room-temperature water into the beaker until the bath level is above the liquid level inside the test tube. Recheck your system for leaks. Now adjust the level of your leveling bulb so that the liquid levels inside the buret and the leveling bulb are as identical as you can make then. This causes the total pressure inside the system to be the same as the outside pressure. Note that the large markings on the buret correspond to milliliters and the smaller markings to tenths of milliliters. Read the height of the water level, *estimating* to the nearest *hundredth* of a milliliter, and record this value in *your laboratory notebook*. (Your instructor will show you how to use the meniscus to obtain reproducible readings. Also see page 34.) The second partner should always verify the buret reading.

Record in your notebook the barometric pressure.

Lift the test tube from the bath and tip it so that a little nitrite solution flows out of the small tube into the acid solution. Effervescence should occur. Return the tube to the bath and shake it gently until the bubbling subsides. Repeat this procedure until all the nitrite solution is mixed with acid. Swirl the tube well to ensure good mixing.

As the gas is evolved, the liquid level in the buret will fall. One partner should move the leveling bulb downward, keeping the two liquid levels about the same as the reaction proceeds. This keeps the pressure from building up inside the apparatus and lessens the chance of popping a connection. (Have a towel or sponge handy and keep the bench top cleared because a fair amount of water will overflow the leveling bulb at this point.)

When the reaction is complete, be sure the bath temperature is still at room temperature and adjust it if necessary. Adjust the leveling bulb to make the liquid levels identical and record the buret reading to the same level of precision as before.

Repeat the above procedure for another sample of pure sodium nitrite.

STUDENTS MUST EACH OBTAIN DATA ON INDIVIDUAL UNKNOWNS.

Next obtain a packet containing a mixture having an unknown percentage of sodium nitrite. *Record the code number of your unknown in your laboratory notebook.* Weigh in your small test tube about the same sample weight of unknown as you used for the pure compound. Be sure to refold your packet carefully to prevent the remaining unknown from absorbing atmospheric moisture. Carry out the procedure as for the pure material. Now repeat with a second sample of your unknown.

Examine your results for consistency. You should find that the sample which weighed more produced the larger volume of gas. If it weighed about 10% more, it should have produced about 10% more gas. If your results are unreasonable, repeat the experiment a third time. You should choose the two runs in which you have the greatest confidence in calculating your percent composition. Be sure to show results of *all* runs in your notebook. If you choose to disregard a run, be sure to explain why.

Treatment of Data

You now have measures of gas volumes, temperatures, and total pressures. The gas you have been trapping in the buret, however, is really a mixture of two gases—nitrogen and water vapor. Hence, the total pressure is really a sum of a partial pressure of nitrogen and a partial pressure of water vapor. A table of the equilibrium vapor pressure of water as a function of temperature is given in the rear of this manual. Subtracting the value of the vapor pressure from the total (atmospheric) pressure gives the pressure of the dry nitrogen.

Assuming ideality, you can now calculate the number of moles of nitrogen formed using the perfect gas equation of state:

$$n = \frac{PV}{RT}$$

The value of R is 82.05 ml atm/mole deg. Although the perfect gas equation of state does not describe the behavior of a real gas precisely, deviations are small under these conditions for nitrogen and are generally less than experimental errors in this experiment.

Suggestions for Organizing Your Notebook

a. Be sure each notebook page you use carries the experiment number and a brief title, your name and desk number, and the date.

b. Open the experiment with your calculation of the weight of $NaNO_2$ needed to produce 40 ml of N_2.

c. Next have a section on known samples. Include as data the name of your lab partner, temperature, pressure, and water vapor pressure, weights of test tubes with and without samples, buret readings. Try to arrange these data intelligently. For instance, set up a table (as indicated in Figure 2-2) so that your weights of test tube with and without sample are already in position for subtraction to obtain the weight of sample.

From these data, calculate the moles of N_2 expected (from the sample weights) and the moles of N_2 observed (from volume of N_2 observed. Use *corrected* pressure and the ideal gas law.). Show at least one calculation of each type you do and label it clearly so your instructor can follow your reasoning. Calculate your percent error according to the formula

$$\text{percent error} = \frac{|\text{moles } N_2 \text{ found} - \text{moles } N_2 \text{ expected}|}{\text{moles } N_2 \text{ expected}} \times 100\%$$

d. Set up a similarly organized section for your unknown sample. In this case, of course, you cannot calculate a "percent error," since you do not have a value for moles of N_2 expected. Instead, you should report the percent of $NaNO_2$ in your unknown (percent purity). This can be obtained by calculating the number of moles of N_2 that would have been obtained if it were pure $NaNO_2$ and comparing that value with the observed moles of N_2 from your sample.

58

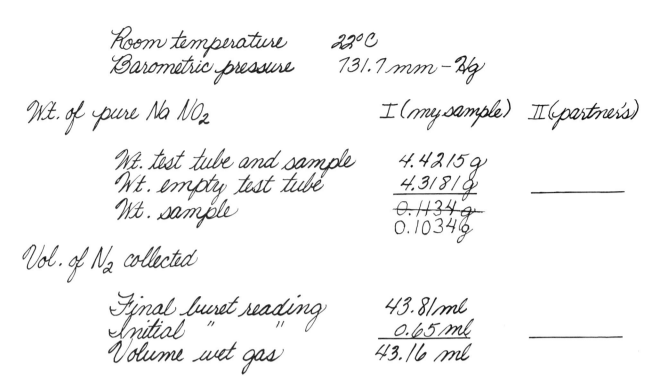

Room temperature   22°C
Barometric pressure   731.7 mm – Hg

Wt. of pure Na NO₂   I (my sample)   II (partner's)

Wt. test tube and sample   4.4215 g
Wt. empty test tube   4.3181 g   _____
Wt. sample   0.1134 g
                      0.1034 g

Vol. of N₂ collected

Final buret reading   43.81 ml
Initial   "   "   0.65 ml   _____
Volume wet gas   43.16 ml

Figure 2-2. Sample Notebook Page, Exp. 2.

e. Conclude your write-up with a discussion and evaluation of likely errors in the experiment. (Note that a blunder such as getting your sample and someone else's switched is not considered an experimental error.) Before you present the actual error analysis for your data, answer the following question. They are suggestive of sources of uncertainty in the experimental measurements.

1. Suppose you made an error of 1 mg in weighing your sample of pure $NaNO_2$. What percent error would this produce in your calculation of number of moles of N_2 expected? Show your work.

2. Suppose your initial and final buret readings were each off by 0.02 ml and that droplets adhering to the walls had a volume of 0.1 ml. Assuming that these errors all worked in the same direction, what percent error would this produce in your calculation of moles N_2 observed? (Did you observe any droplets adhering to the walls of your buret?)

3. Suppose your sample of $NaNO_2$ was only 98% pure, the remainder being absorbed moisture or other inert impurities. How would this affect your moles N_2 expected?

4. Suppose you made an error of 2°C in the temperature of the system. This can affect your results in two ways, one direct and one indirect. What are these two ways (how does the temperature affect the values used in the calculation)? For each effect, calculate the percent error produced by the 2° temperature error, and determine the total error this will give in your calculation of moles of N_2 observed.

5. Suppose your gas sample was only 90% saturated with water vapor. What is the percent error this would cause in your calculation of number of moles of N_2 collected? Show your work. [NOTE: This question gives you an idea of the importance of a lack of saturation in affecting your result. Actually in an experiment, there is no way you can estimate whether the saturation is 50%, 90%, or 100%. Thus, it is inappropriate to include this factor in a formal estimate of your experimental error.]

On the basis of consideration of the uncertainties in the various experimental values you measured, prepare an analysis of the total ± uncertainty in *both* the moles of N_2 expected and the moles of N_2 found. The analysis can be summarized in tables like the following:

Moles N_2 expected (or found)

	Value	*Est. Uncertainty*	*% Error*
A			
B			
C			

where A,B,C. . . are the measurements used to calculate the number of moles. [What one experimental measurement fixes the number of moles *expected*? By contrast, there are three measurements which affect the number of moles *found*, with one of these entering two ways.]

For each of the quantities A,B,C. . . make realistic estimates of the uncertainties for the manner in which you performed the experiment.* Finally, from the errors in the individual quantities, compute the expected maximum uncertainty in the number of moles, found and expected. Do your actual results on pure $NaNO_2$ agree within these limits? If not, can you think of any blunders or mistakes that could be responsible for the disagreement? Use this analysis to set ± values for the reported percent of $NaNO_2$ in your unknown.

Questions

Be sure you can answer these when you come to lab. There may be a quiz.

1. Why must the water in the gas generator tube and buret be at room temperature?
2. Why must the water levels in the leveling bulb and buret be identical whenever a reading is taken?
3. Why must the pre-dried $NaNO_2$ be weighed without delay after you obtain it?
4. What is the equation for the chemical reaction you are performing?
5. What weight of $NaNO_2$ would be required to produce 40 ml of dry N_2 gas at standard pressure and 25°C according to equation (2)?
6. Suppose you can measure a gas volume to ± 0.10 ml. What will be the percentage errors of measurement in a volume of 40.00 ml and in one of 8.50 ml? Does this clarify why you use enough $NaNO_2$ to yield ∿40 ml of gas (see top of p. 56)?
7. Suppose you failed to level the bulbs as described in Question 2 above. Further, suppose the two *water* levels differed by 0.5 cm. What error (in Torr) would this produce in your determination of the pressure? (The density of water is 1.0 g cm⁻³, that of liquid Hg is 13.6 g cm⁻³.)

NOTE: No solid $NaNO_2$ or unknowns should be thrown into the trash cans. Wash the soluble solids down the drain. ($NaNO_2$ is destroyed in the sewer plant.)

---

*A subtle point enters in consideration of the uncertainty in the temperature. You might reasonably say you could read the thermometer to ± 0.2°C. However, the student thermometers are uncalibrated and may be "off" in their readings. A guess to their accuracy can be obtained by inter-comparing several when they are all put into the same beaker of water. This will indicate the variation in calibration and give a better estimate of the uncertainty in your temperature measurement.

Synthesis and Analysis of a Coordination Compound; Redox Volumetric Analysis

The purpose of this experiment is to synthesize a complex salt and then analyze it to determine its simplest formula, exemplifying the laws of stoichiometry.

PART I: Synthesis of a Complex Salt of Iron

Theory

A complex salt is basically an ionic compound (hence, a "salt") but it differs from simple salts in that the metal cation is associated with a number of ions or molecules (called "ligands") via **coordinate covalent bonds** (i.e., *both* bonding electrons come from the ligand and the metal cation provides empty d-orbitals). The water molecule itself can act as a ligand, and many ions in solution are really complexed by water molecules. Cupric ions are one example. In solution they form a complex with a strong blue color. Upon evaporation of the solution at room temperature, blue crystals of the complex salt (e.g., copper sulfate—$CuSO_4 \cdot 5\ H_2O$) are formed. (The formula would be more indicative of the true situation if it were written $Cu(H_2O)_4(SO_4) \cdot H_2O$, since four of the water molecules are bound directly to the copper ion.) Upon strong heating of these crystals, the water is driven off and a simple white salt, $CuSO_4$, is formed.

In this experiment, you will prepare a complex salt of iron in which water molecules and oxalate ions ($C_2O_4^=$) act as ligands. The formula for the complex will be of the form $K_x Fe_y (C_2O_4)_z \cdot 3\ H_2O$. After preparing the salt, you will analyze it for its percentage composition, and from this determine the values for x, y, and z in the above formula.

Throughout this experiment you will be utilizing the fact that iron can exist in solution in two oxidation states. These are the ferrous (Fe^{2+}) and ferric (Fe^{3+}) states. These ions can be made to interconvert by changing the oxidizing-reducing nature of their surroundings. Thus, if a good reducing agent (e.g., Sn^{++}) is added to a solution containing Fe^{3+} ions, the stannous ions become oxidized to stannic ions and the ferric ions are reduced to ferrous ions:

$$2\ Fe^{3+} + Sn^{2+} \rightarrow 2\ Fe^{2+} + Sn^{4+}. \tag{1}$$

The oxidation of ferrous ions to ferric ions can be effected by an oxidizing agent like hydrogen peroxide:

$$2\ H^+ + 2\ Fe^{2+} + H_2O_2 \rightarrow 2\ H_2O + 2\ Fe^{3+} \tag{2}$$

Even oxygen from the air can oxidize ferrous ions in acid solution:

$$O_2 + 4\ Fe^{2+} + 4\ H^+ \rightarrow 4\ Fe^{3+} + 2\ H_2O. \tag{3}$$

In the synthesis of the complex salt, you will dissolve a ferrous compound and form the precipitate ferrous oxalate:

$$Fe^{2+} + C_2O_4^{2-} \rightarrow FeC_2O_4 \downarrow. \tag{4}$$

61

After washing this precipitate, you will oxidize the iron in the presence of excess oxalate ions by adding hydrogen peroxide:

$$\underset{(2+)}{K^+ + FeC_2O_4} + H_2O_2 + C_2O_4^{2-} = \rightarrow \underset{(3+)}{K_x Fe_y(C_2O_4)_z} \cdot 3H_2O + OH^- \qquad (5)$$
$$\text{(not balanced)}$$

After allowing the desired product to crystallize, you will wash it and proceed with the analysis described in later sections of this experiment.

Experimental Procedure

Every student should work individually on this entire experiment.

NOTE: You must complete the synthesis of the complex in the first period so the solid can crystallize between lab meetings.

Be sure to enter all weights and observations about colors, color changes, precipitates, etc. into your notebook. It would be sensible to label each section of your notebook to agree with the relevant section of the write-up. That is, this first section could be labelled "I. Synthesis of a Complex Salt of Iron."

Weigh out about 10 g of ferrous ammonium sulfate hexahydrate, $Fe(NH_4)_2(SO_4)_2 \cdot 6H_2O$. (Use the triple beam balance and record the weight of your sample to the nearest 0.1 g.) In a 250-ml beaker, add 15 drops of 3M H_2SO_4 (in glass-stoppered bottle on shelf over desks) to about 30 ml of distilled water. Warm the solution a little and then dissolve the sample of ferrous ammonium sulfate. While stirring this solution, add 50 ml (graduated cylinder) of 1M oxalic acid ($H_2C_2O_4$).

A precipitation reaction according to equation 4 will occur. *Carefully* heat the mixture *nearly* to boiling while stirring. Let it cool and settle. The liquid layer above the precipitate should be clear. Gently pour off (decant) as much of the liquid as possible, discarding the liquid and retaining the solid precipitate. Then remove as much additional liquid as possible by the use of an eyedropper. To wash the precipitate, add about 30 ml of almost boiling distilled water to the precipitate in the beaker, stir, and allow the precipitate to resettle. Repeat the decantation of the wash liquid as before. Finally, wash the precipitate in this way once more.

To the beaker containing the washed FeC_2O_4 precipitate, add 18 ml (graduated cylinder) of saturated (ca. 1.8 M) potassium oxalate ($K_2C_2O_4$) solution. Heat this mixture to 40°-45°C and place the beaker inside a larger beaker containing water at this temperature so that the temperature will be maintained. Add *slowly, from a buret,* 17 ml of 6% hydrogen peroxide (H_2O_2) stirring continuously. (This should only take a couple of minutes.) The solution may become basic enough to precipitate some ferric hydroxide, $Fe(OH)_3$, (see equation 5) but this will be redissolved in the subsequent step. Remove the beaker from the warm water bath and heat it until the mixture is almost boiling. Then add 9 ml (graduated cylinder) of 1M $H_2C_2O_4$ all at once. Continue adding 1M $H_2C_2O_4$ dropwise, while stirring, until you have a clear solution. (If you still have some precipitate after you have added a total of 15 ml of $H_2C_2O_4$, filter the hot solution and discard the precipitate.) Add to the solution about 20 ml of 95% ethyl alcohol (C_2H_5OH). (This prevents the precipitation of crystals of $H_2C_2O_4$, which is more soluble in alcohol-water than in pure water.) If some crystallization has already occurred, heat the solution gently UNDER THE HOOD (alcohol is flammable) until all crystals are redissolved. Place your beaker in your desk and cover it with an inverted beaker of larger size or else with a piece of aluminum foil. Allow it to remain undisturbed until the next laboratory period.

At the next laboratory period, you should find a mass of green crystals covering the bottom and sides of the beaker. Prepare a suction filter as demonstrated by your instructor. Using a rubber policeman, scrape your crystals and solution into the funnel. If any crystals remain in the beaker, wash then into the funnel with some of the filtered liquid (called mother liquor) from the filter flask. Rinse the crystals in the funnel with 15 ml of a 1:1 alcohol-water solution, poured slowly so all of the crystals are washed.

To dry your crystals rapidly, pour three 10-ml portions of acetone over them and then allow them to dry in the air. (This should not take more than 15-20 minutes.) NOTE: Acetone is flammable and highly volatile. **Do not use it in the presence of open flames.** (No student should use a burner during the first part of the second lab meeting.) Alcohol-water-acetone wastes should be disposed of in the lab sinks accompanied by a thorough flushing with water **as soon as you are through with them, and before any burners are lit.**

When the crystals are *thoroughly dry,* transfer them to a previously weighed weighing bottle (triple beam balance). Reweigh the bottle to obtain the weight of the crystals. Since these crystals are decomposed by light, wrap your weighing bottle in aluminum foil when it is not in use.

Questions

Be sure you can answer these when you come to lab. There may be a quiz.

1. In synthesizing the complex salt, what is the reason for adding the hydrogen peroxide?
2. What is meant by decanting a liquid?
3. What precaution must be followed in the use of alcohol in the laboratory?
4. The boiling points of ethyl alcohol and of acetone are 79 and 56 C., respectively. Which is the more flammable substance?
5. What two uses is ethyl alcohol put to in the preparation and isolation of the complex salt?
6. What precautions must be taken in the addition of the hydrogen peroxide?

PART II: Analysis of the Complex Salt

In the remainder of this experiment, you will determine the percentage composition of the complex salt. To simplify this determination, we will assume the salt to be a trihydrate (contains three water molecules per simplest formula). The amount of iron and oxalate ions in a known amount of the salt will be determined as described below.

A. Standardization of Potassium Permanganate Solution

Theory

The determinations of oxalate and iron involve oxidation-reduction ("redox") titrations. (The technique of titration will be demonstrated for you.) In both cases, the **titrant** is a solution of potassium permangante ($KMnO_4$) of known concentration. The first task, then, is to make an accurate measurement of the concentration of a solution of potassium permanganate. You will do this by titrating the $KMnO_4$ solution into a solution containing a known weight of sodium oxalate ($Na_2C_2O_4$). The following reaction occurs:

$$5 C_2O_4{}^{2-} + 2 MnO_4{}^- + 16 H^+ \rightarrow 10 CO_2 + 2 Mn^{2+} + 8 H_2O \qquad (6)$$

The permanganate ion ($MnO_4{}^-$) has a strong violet color whereas all the products are colorless. Hence the solution in the flask will remain colorless until all the oxalate ion is oxidized. The first trace of *persistent* violet coloration in the flask marks the **endpoint** of the titration. You must swirl the flask as the titration is performed. (Your instructor will demonstrate the proper technique for swirling a liquid in an Erlenmeyer flask.) If a high local concentration of permanganate is allowed to collect, the permanganate may be only partially reduced:

$$2 MnO_4{}^- + 3 C_2O_4{}^{2-} + 8 H^+ \rightarrow 6 CO_2 + 2 MnO_2\downarrow + 4 H_2O. \qquad (7)$$

This situation is signaled by a brownish coloration due to suspended manganese dioxide (MnO_2). If you have not added more $KMnO_4$ than needed to reach the endpoint, the excess remaining oxalate ion should reduce the MnO_2 produced momentarily as follows:

$$C_2O_4^= + MnO_2 + 4\,H^+ \rightarrow 2\,CO_2 + Mn^{2+} + 2\,H_2O \qquad (7a)$$

Thus the overall $MnO_4^-/C_2O_4^=$ stoichiometry is preserved. If, however, you fail to swirl the sample and overshoot the endpoint while MnO_2 is formed via equation 7, the titration is ruined and must be discarded.

A properly titrated solution will be faintly pink at the endpoint. Upon standing for several minutes, it will likely develop a brown cloudiness as the excess MnO_4^- ion is reduced by the Mn^{++} produced during the reaction with oxalate ion (equation 6). This does not invalidate the titration.

Experimental Procedure

NOTE: Since impurities are frequently oxidized by permanganate ions, it is important that all apparatus be clean and that distilled water be used in all solutions.

First, clean your polyethylene bottle of any MnO_2 (dark brown) deposits. Do this by reducing the Mn(IV) to Mn^{2+}; add 1-2 ml of 6 M HCl and an equal volume of H_2O_2 solution. Cap and shake the bottle so the solution comes in contact with all the surfaces. Rinse well.

Now rinse your bottle with two 2-ml portions of the permanganate stock solution to be standardized, and then take about 250 ml of the solution. This solution has an *approximate* concentration of 0.02 M. Stopper this bottle when it is not in use.

In much of the rest of this experiment, you will need to use distilled water for such purposes as washing down the walls of titration vessels. Your instructor will demonstrate the technique of using your wash bottle (polyethylene squeeze bottle) to provide a stream for this and other chores. Clean and rinse your wash bottle and fill it with distilled water (after rinsing three times with 5-ml portions). Once filled, the wash bottle need not be emptied at the end of a lab period.

Weigh out (to the nearest 0.1 mg) a sample of sodium oxalate weighing approximately 0.12 g in a 250-ml Erlenmeyer flask. Prepare approximately 200 ml of 0.75 M H_2SO_4 by adding 10 ml of *concentrated* reagent-grade H_2SO_4 to 200 ml of distilled water. (Never add water to concentrated H_2SO_4, as it tends to spatter explosively.) The concentrated sulfuric acid is in a screw-cap bottle on the shelf over your desk. Add 75 ml of the dilute solution to the flask to dissolve your sample of sodium oxalate. Save the rest of the acid for use as a titration blank (see below).

Reread the instructions for cleaning and rinsing a buret (page 34). After rinsing the buret with three small (3-5 ml) portions of $KMnO_4$, fill it to slightly above the zero marking. Let enough drain rapidly out the tip so that you can read the graduations at the top of the liquid column. This should also sweep any air bubbles out of the tip. Remove any drop adhering to the tip by rinsing it with a fine stream from your wash bottle. Note that it is possible to estimate the position of the top of the liquid column to two decimal places (e.g., 0.32). Since the solution is very dark, you should take readings from the top edge of the liquid rather than from the bottom of the meniscus.

Heat the solution of sodium oxalate in the Erlenmeyer flask to 80-90°C. Remove your thermometer from the solution prior to titrating, but rinse it off *into* the flask as you are withdrawing it so as not to carry away any of the sodium oxalate. Record the initial reading of the buret to two decimal places in your laboratory notebook and immediately begin to add permanganate solution to the flask. Do not add it too rapidly. (About 5 drops per second will do.) Swirl the solution. If you add $KMnO_4$ too rapidly, or do not swirl sufficiently, or if the flask is not hot enough, you may get some brown MnO_2 (equation 7). You will notice that the purple titrant is being decolorized as it falls into the hot sodium oxalate solution, and that the decoloration takes longer and longer. As the time for decoloration increases, let the titrant fall in more slowly (drop-by-drop). Finally, the addition of a single drop of titrant will cause a faint pink color to persist in the solution. (Placing a sheet of white paper under your flask makes the end point more visible. Having a flask of water available for color comparison is also helpful.) Record the final reading of the buret, again to two decimals.

Since the sulfuric acid solution itself may contain some impurities which react with permanganate, it is useful to run a blank. Without a correction which this will provide, the data would contain a systematic error. Add about 75 ml of the 0.75 M H_2SO_4 to a clean Erlenmeyer flask and heat it up to 80-90°C. You may use the final reading from your last titration as the initial reading for this one. Titrate until a persistent (4-5 seconds) pink coloration occurs. This should take only one or two drops. Record the final reading.

Treatment of Data

To obtain the volume of $KMnO_4$ solution needed to oxidize the oxalate ion, subtract the volume of the blank from the volume used in the sample. Since you know the number of moles of sodium oxalate used in your run, you can use equation 6 to calculate the number of moles of permanganate ion used in the titration. This, together with the volume of solution, gives the molarity of the $KMnO_4$ solution.

Your instructor will collect the molarities of $KMnO_4$ determined by your section and list them on the board. Record these in your notebook. After striking out any values which are obviously in error, determine the class average value. You should take this average value as the concentration of the $KMnO_4$ solution.

Organization of Notebook

Your notebook-lab report should be completed section-by-section as you perform this experiment. For this part of the experiment, it will be convenient to divide your record according to the following items, all to be done before you proceed to Part B.

    a.   weight of $Na_2C_2O_4$ sample
    b.   moles $Na_2C_2O_4$
    c.   ml MnO_4^- titrated into sample
    d.   blank titration
    e.   calculation of molarity of MnO_4^-
    f.   calculation of class average value. Estimation of error limits on molarity.

Don't forget to arrange your data in an order that makes for natural subtraction (e.g., "final volume" above "initial volume"). Pay attention to significant figures. Estimate your precision and compare this estimate with the scatter in class values for $KMnO_4$ molarity. Aside from those cases where a blunder has obviously been made, how well does the precision evidenced by the class data accord with your estimate based on the capabilities of your apparatus? If there is a discrepancy, can you cite factors which you think might be responsible? If you differ from the class average value by more than the precision range you expected, can you think of any likely causes? Consideration of these things now might prevent you from repeating mistakes in the remainder of this experiment.

As an aid towards organizing your notebook, a sample page for this section is given in Figure 3-1.

B. Analysis for Oxalate in the Complex Salt

Theory

We have just used a known amount of oxalate ion to standardize an unknown solution of $KMnO_4$. Now we will simply reverse the process and use the standardized solution of $KMnO_4$ to determine an unknown amount of oxalate.

Jane Doe
Experiment 3 (cont'd)
March 28, 1978

II.A. Standardization of $KMnO_4$ soln.

$Na_2C_2O_4$ sample		$KMnO_4$ soln
Wt. of sample:		Titrated into $Na_2C_2O_4$ sample
Wt. flask + $Na_2C_2O_4$ _____ g		Final buret reading _____ ml
Wt. flask _____ g		Initial " " _____ ml
Wt. sample _____ g		Vol. $KMnO_4$ _____ ml
Moles of $Na_2C_2O_4$		Blank Titration
	2 Na 45.98	Final buret reading _____ ml
	2 C 24.00	Initial " " _____ ml
	4 O 64.00	Vol. of blank _____ ml
	133.98	
Moles $Na_2C_2O_4$ =		Corrected vol.: _____ ml
		_____ ml (blank)
$\dfrac{\underline{\qquad} g}{133.98\ g/mole}$ = _____ moles		_____ ml

Molarity of $KMnO_4$

Balanced equation: $5 C_2O_4^{2-} + 2 MnO_4^- + 16 H^+ \longrightarrow 10 CO_2 + 2 Mn^{2+} + 8 H_2O$

$$5\ moles\ C_2O_4^{2-} = 2\ moles\ MnO_4^-$$

Moles MnO_4^- = (_____ moles $C_2O_4^{2-}$) × $\dfrac{2\ moles\ MnO_4^-}{5\ moles\ C_2O_4^-}$ = 0.4 (_____) moles MnO_4^-

Molarity $KMnO_4$ soln. = $\dfrac{moles\ MnO_4^-}{l\ soln}$ = _____ M.

Figure 3-1. Sample Notebook Page, Part IIA.

Experimental Procedure

Weigh (to 0.1 mg) two 0.10-0.15 g samples of your complex salt in two, labelled 250-ml Erlenmeyer flasks. Dissolve the samples in a few ml of 3 M H_2SO_4 (dilute acid, on shelf over desks in a glass stoppered bottle). If necessary, warm to dissolve the sample. (Note the number of ml used so that you will know how to prepare a blank.) Dilute each sample to about 75 ml. Heat to $80\text{-}90°C$ and titrate as you did in the standardization. The endpoint color change may be somewhat different than before due to the presence of ferric ions. A blank should be run on a sample prepared identically as above but without any complex salt.

Treatment of Data

Since you know the volume and molarity of the $KMnO_4$ solution, you can calculate how many moles of $KMnO_4$ were used. This leads to the number of moles of $C_2O_4^{2-}$ present from equation 6. Conversion to grams and comparison with the total weight of the sample gives percent oxalate. Calculate an average percent oxalate in the salt based on your two determinations. Show all data and calculations in a well-organized format in your notebook.

C. Analysis for Iron in the Complex Salt

Theory

The iron in the complex salt is in the *ferric* (Fe^{3+}) state. You will first reduce it to the *ferrous* (Fe^{2+}) state and then titrate it with your standardized $KMnO_4$ solution to re-oxidize it:

$$5\ Fe^{2+} + MnO_4^- + 8\ H^+ \rightarrow 5\ Fe^{3+} + Mn^{2+} + 4\ H_2O. \qquad (8)$$

There are certain complications, however. Since oxalate ions also react with permanganate, we must somehow get rid of all the oxalate ions before titrating. This is done by first adding a strong solution of $KMnO_4$ to the dissolved salt. This permanently removes oxalate as CO_2 gas as shown in equation 6. Next, we must reduce the iron to the ferrous state. This is done by adding excess stannous ions, causing the redox reaction.

$$Sn^{2+} + 2\ Fe^{3+} \rightarrow Sn^{4+} + 2\ Fe^{2+} \qquad (9)$$

The stannous ions also react with and remove any excess permanganate left over from the previous step.

Now the oxalate ions are gone and the iron is reduced. But we cannot titrate yet because the remaining Sn^{2+} ions would also react with the titrant. So we add a quantity of mercuric (Hg^{2+}) ion which causes the reaction

$$2\ Hg^{2+} + Sn^{2+} + 2\ Cl^- \rightarrow Hg_2Cl_2 \downarrow + Sn^{4+}. \qquad (10)$$

This leaves us with a solution containing ferrous ions but no other ion which will be oxidized by MnO_4^-. The following "flow scheme" shows the ions of interest in solution after each addition.

$$\frac{Fe^{3+}}{C_2O_4^=} \xrightarrow[\text{eq'n 6}]{\overset{CO_2\uparrow}{[\text{excess } MnO_4^-]}} \frac{Fe^{3+}}{MnO_4^-} \xrightarrow[\text{eq'n 9}]{[\text{excess } Sn^{2+}]} \begin{matrix} Fe^{2+} \\ Mn^{2+} \\ Sn^{4+} \\ Sn^{2+} \end{matrix} \xrightarrow[\text{eq'n 10}]{[\text{excess } Hg^{2+}]} \begin{matrix} Fe^{2+} \\ Mn^{2+} \\ Sn^{4+} \\ Hg^{2+} \\ Hg_2Cl_2 \downarrow \end{matrix}$$

67

You are much less likely to make an experimental blunder if you understand the reason for each step in this analysis. Be sure that you understand this section before attempting to do the experiment.

Experimental Procedure

Weigh accurately (to the nearest 0.1 mg) two samples weighing between 0.5 and 0.7 g of the complex salt into two labelled 250-ml Erlenmeyer flasks. To one of your samples, add about 10 ml of 3 M H_2SO_4. Warm to dissolve the sample.

You will now remove the oxalate ions from the solution. While the solution is warm, add 3% $KMnO_4$ (NOTE: This is not the standardized solution, but a separate reagent on the side shelf.) dropwise from a pipet until the first permanent appearance of a pink coloration or brown precipitate (what could it be?). You may observe effervescence of the solution (explain).

You will now reduce the iron and any excess permanganate. The permanganate is reduced first. Add stannous chloride ($SnCl_2$) solution carefully from a dropper until a clear yellow solution results. (Yellow due to Fe^{3+}). Any precipitate should redissolve. ($MnO_2 \rightarrow Mn^{2+}$) Carefully bring the solution to a low boil and continue the dropwise addition of $SnCl_2$ until the yellow color of the ferric ions has just disappeared and then add *1 or 2* drops more. Only about 40 drops (2 ml) should be required.

You will now oxidize the left-over stannous ions. Cool the solution under the cold water tap or in a cold water bath and add 10 ml (graduated cylinder) of mercuric chloride ($HgCl_2$) solution *all at once*. You should observe a white, silky precipitate of mercurous chloride (Hg_2Cl_2). If no precipitate results, you failed to add sufficient $SnCl_2$ in the previous step. If a gray precipitate results, you have produced elemental mercury due to too great an excess of stannous chloride or too slow addition of mercuric chloride:

$$Hg^{2+} + Sn^{2+} \rightarrow Hg + Sn^{4+} \tag{11}$$

In either event, you must begin your iron determination again.

You are now ready to determine the ferrous ion concentration by titration with permanganate. But, since ferrous ions tend to revert to ferric ions in the presence of air (equation 3), you must proceed without delay. Dilute the solution by adding to it approximately 100 ml of distilled water and 25 ml of Zimmerman-Reinhart reagent. (This reagent prevents oxidation of chloride ions by permanganate in the titration of iron. It also ties up ferric ions as the phosphate complex, shifting the equilibrium of reaction 8 to the right and making the oxidation more complete.) Titrate slowly with your *standard* permanganate solution to a faint pink coloration which persists for 4 to 5 seconds.

Repeat this procedure on your second sample.

Carry through the above procedure, omitting the sample. Add only one drop of 3% $KMnO_4$ solution and only 1-2 drops of $SnCl_2$ as there are no oxalate or ferric ions present. Use this solution to run a blank.

Treatment of Data

Knowing the volume and molarity of the $KMnO_4$ solution enables you to calculate moles of MnO_4^- used. This, with equation 8, gives the number of moles of iron present. From this you can calculate grams of iron and percent iron in the sample.

Organization of Notebook

In Parts B and C you have obtained the percentage composition of the salt with respect to oxalate and ferric ions. These can now be combined to solve the problem of the formula of the compound posed at the beginning of this experiment.

Calculation of the Empirical Formula of the Complex. You first must calculate the number ratio (same as the mole ratio) of oxalate ions to ferric ions in the complex. This is done by dividing the percent by the molecular weight of the species of interest. For example, a certain compound is 27.25% C, 72.75% O by weight.

Taking 100 g of the compound and dividing each element by its atomic weight gives:

$$C: \frac{27.25 \text{ g C}}{12.00 \text{ g C/mole C}} = 2.275 \text{ mole C} \qquad O: \frac{72.75 \text{ g O}}{16.00 \text{ g O/mole O}} = 4.550 \text{ mole O}$$

$$\frac{\text{mole C}}{\text{mole O}} = \frac{2.275 \text{ mole C}}{4.550 \text{ mole O}} = 0.500 \frac{\text{mole C}}{\text{mole O}}$$

But the mole ratio is the same as the ratio of number of atoms. Thus, there are 2 oxygen atoms for each carbon atom and the simplest, or empirical, formula for this compound is CO_2. The true formula might be CO_2, C_2O_4, C_3O_6, etc. We cannot distinguish among these on the basis of percent composition.

For further discussion of formula determination see Brown and LeMay, p. 77; Masterson and Slowinski, Ch. 3; Mortimer, pp. 142-143; Sienko and Plane, pp. 129-133.

Using this approach, find the ratio of ferric to oxalate ions. This ratio should be quite near an integer or simple fraction. Round off to the nearest likely value, and use this to construct likely formulas for the complex (assume a trihydrate). Use potassium ions to yield a *neutral* formula. For example, if your mole ratio were 1.48, you would round off to 1.5 = 3/2, or three oxalates for every two irons. Possible formulas would be

$$Fe_2(C_2O_4)_3 \cdot 3 H_2O, \ Fe_4(C_2O_4)_6 \cdot 3 H_2O, \text{ etc.},$$

(no potassium ions are required to balance charge here). For each of your postulated formulas, calculate the percent composition of oxalate and iron and compare with your experimental values. The formula which gives best agreement is your experimentally determined "simplest formula." All this work should appear in your notebook in a neat format.

Once you know the simplest formula, you can figure out the theoretical maximum weight of crystals you could have produced from ten grams of ferrous ammonium sulfate hexahydrate. Compare this to the actual weight of your product by calculating the percent yield:

$$\% \text{ yield} = \frac{\text{actual weight produced}}{\text{theoretical maximum weight}} \times 100.$$

Error Analysis. The ultimate result of your analyses was the ratio of moles of oxalate to moles of iron (the number ratio), which fixed the formula. The ratio was based on the individual analyses, which in turn depended upon the concentration of the $KMnO_4$ used for titration.

Discuss the precision of your analysis for iron and for oxalate, based on the uncertainties in your titration, your weighings, and any other factor(s) that must be considered. Compare your precision as measured by the spread of values for your duplicate runs with the expected U_m.

Calculate the error limits for your number ratio, based on the uncertainties in your two percentages. Comment on whether the integral number ratio you selected agrees with your raw experimental value within the error limits.

Include an answer to the following question. What effect on your number ratio would result if your acetone rinsing had not removed all the moisture from the complex so that the sample still contained 0.5% water (by weight)?

Questions

Be sure you can answer these when you come to lab. There may be a quiz.

1. Why must the flask be swirled continuously during a redox titration with permanganate ions?
2. Why is no indicator added in any of the titrations in this experiment?
3. What is the characteristic color of ferric ions in solution?
4. Why is a solution of $KMnO_4$ added to the complex salt solution *prior* to carrying out the redox titration for iron content?
5. Why is stannous chloride solution added in the iron determination?
6. Why is mercuric chloride solution added in the iron determination?
7. In a titration using $KMnO_4$ solution, the following data were obtained:

Buret readings   FINAL   15.00 ml   Precision of reading ± 0.10 ml

INITIAL   5.00 ml

Labelled concentration of $KMnO_4$ 0.0200 ± 0.0003 M

How many moles of $KMnO_4$ were used? (Include the uncertainty in the answer.)

Liquid-Vapor Equilibrium; Thermodynamics of Vaporization

The purpose of this experiment is to determine the vapor pressure of an organic liquid at several temperatures and, from this, calculate the heat of vaporization.

Theory

Imagine a closed bulb containing only liquid water and water vapor. At low temperatures, relatively few molecules will have sufficient kinetic energy to overcome the attractive intermolecular forces which stabilize the liquid phase. Hence the vapor pressure will be low. As the temperature increases, more molecules will become sufficiently energetic to break away from the liquid and the vapor pressure will rise. At $100°C$, the vapor pressure will be equal to 1 atmosphere. In the closed bulb, continued heating will raise the temperature and vapor pressure still higher, just as in a pressure cooker. In an open system, it is impossible to maintain a pressure higher than the prevailing barometric pressure, so the temperature will stop rising somewhere around the boiling point. The water vapor bubble formation characteristic of boiling occurs when the vapor pressure of the water is high enough to equal the pressure of the atmosphere-plus-water which acts to collapse the bubbles.

The **Clausius-Clapeyron equation** relates the vapor pressure of a liquid to its temperature:

$$\log_{10} P = \frac{-\Delta H_v}{2.303\,R}\left(\frac{1}{T}\right) + I.$$

P is the vapor pressure in mm of Hg, T is the temperature in degrees Kelvin, ΔH_v is the heat of vaporization of the liquid (the amount of heat needed to convert one mole of liquid to one mole of vapor at the same temperature and pressure), and R is the gas constant (8.3143 Joules/deg. mole). I will be taken as constant and will cancel whenever the above equation is used to give *changes* in pressure with changes in temperature.

In this experiment, you will determine the vapor pressure of an unknown organic liquid at several temperatures. The closed end of your **isoteniscope** (made in the glassworking exercise) provides a chamber in which organic vapor and liquid will be kept. The isoteniscope will be submerged in a water bath so you can control the temperature of both phases. You will adjust the pressure on the open end of the isoteniscope until the liquid levels in the two arms are equal. This occurs only when the pressure of the trapped organic vapor is equal to the pressure on the open end of the isoteniscope. Hence, by measuring the pressure on the open end of the isoteniscope, you will, in effect, measure the vapor pressure of the organic compound at the given temperature. While the isoteniscope itself is a simple device, the pressure controlling and measuring apparatus is fairly complicated in appearance. However, it is not difficult to understand this apparatus once you understand its purpose—control and measurement of pressure. This apparatus will be described in the EXPERIMENTAL section below.

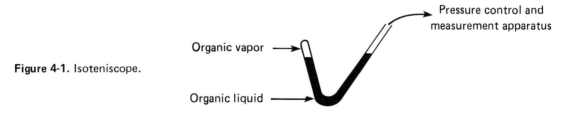

Figure 4-1. Isoteniscope.

Experimental Procedure

Students should work in pairs on this experiment. Submit separate reports on the single unknown used by the pair.

Submit to your instructor a clean, dry, labelled test tube. This will be returned to you with a sample of your unknown organic liquid. Keep the test tube stoppered so your unknown doesn't evaporate.

The stockroom will provide an assembled manifold made of glass tubing and tygon tubing. Please notice how many pieces are involved in this manifold so you can leave it, **in its entirety**, on your desk at the end of the day. You will also be provided with one long piece of heavy walled rubber tubing ("vacuum" tubing), and a closed-end mercury manometer mounted on a wooden frame.

The manometer is a fragile piece of equipment. Rapid pressure surges can drive the mercury column against the closed end with enough force to break the glass. Do not play with spilled mercury. Mercury vapor is poisonous, and its effects are cumulative over a lifetime. **If manometer breakage or mercury spillage occurs, report it to your instructor at once.**

The closed end of the manometer has a near-vacuum trapped over the mercury column. The other end of the manometer is open to the atmosphere. Hence we have a mercury column plus zero gas pressure on the closed side balancing a shorter mercury column plus atmospheric pressure on the other. It follows that the *extra height of mercury in the closed side balances the atmospheric pressure on the open side.* Thus, the *difference* in mercury column heights provides a measure of atmospheric pressure in mm of mercury. (The pressure exerted by a 1-mm column of mercury is called a "torr" after Torricelli, who invented the barometer in 1643.) If we attach the open end of the manometer to a closed system and vary the pressure in that system, the mercury columns will move in such a way that their height *difference* corresponds to the pressure in the system. If we were to remove all the gas from the system, the mercury columns would become equal in height since we would then have mercury plus zero pressure balancing mercury plus zero pressure.

At the beginning of the period, read the two column heights in your manometer, and compare their difference with the pressure registered by the laboratory barometer. Remember that these readings, like all data observed in the laboratory, should be recorded in your notebook when they are made. If there is disagreement of 5 torr or more, the vacuum in the closed end of the manometer is not good enough. In this case, notify your instructor so he can tell you how to proceed.

You will assemble the pressure control and measurement apparatus using the above mentioned components and equipment from your desks. The apparatus is sketched in Figure 4-2. Stopcocks A and B should be taken from your burets. Notice that the glass tips are not to be removed from the stopcocks. Use pieces of tygon tubing to fasten the stopcocks to the apparatus.

It will be necessary to suspend the ballast bottle and trap bottle by clamping. Be sure to use glycerine when inserting glass into rubber. **Do not force the glass!** Connect your buret stopcocks as shown in the sketch, using short pieces of tygon tubing.

This pressure control and measurement apparatus looks complicated, but it is not difficult to understand. The manometer measures the pressure in the apparatus, the pressure being given by the difference in column heights of mercury. The aspirator on the water faucet is a simple pump which draws gas out of the apparatus. The trap bottle catches any water that "backs-up" into the system as the water flow rate varies at the faucet. (If the water flows rapidly at first, it will produce a low pressure in the apparatus. If someone else turns on another faucet, the flow rate through your faucet may decrease. Then the pressure in your apparatus may be lower than that which can be maintained by your aspirator and the apparatus will "suck" water out of the aspirator.) The ballast bottle simply makes the volume of the system larger, so that, as you let air in or out of the system, the pressure will change fairly slowly, giving you better control.

There are two openings into the system, and these are controlled by two buret stopcocks (labelled A and B in the figure). When you wish to remove gas from the system, turn on the water

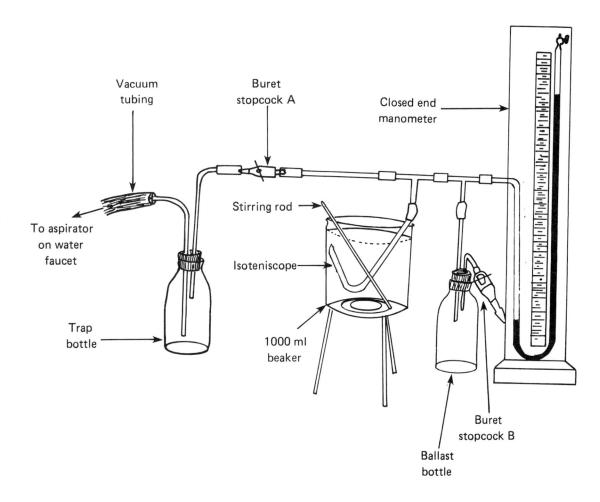

Figure 4-2. Apparatus for Vapor Pressure Measurement.

faucet, open buret stopcock A and close buret stopcock B. When you have evacuated sufficiently, close stopcock A. **Do not turn off the water tap until you have closed stopcock A.** To let air into the system, open stopcock B. Air will then bleed in slowly through the capillary. The rate of flow can be controlled by the extent to which the stopcock is opened and by the use of your finger on the opening of the glass tip. Thus, you have an apparatus which enables you to control and measure the pressure on the open end of your isoteniscope.

In some laboratories, it is necessary to connect more than one apparatus to the same aspirator through a T-joint. (This is necessary since, if all the faucets are on simultaneously, the water pressure drops too low to give effective pumping action.) If your apparatus shares an aspirator with another setup, you must put a screw clamp on your vacuum tubing and **be sure that your tubing is clamped shut whenever the other apparatus is open to the aspirator,** and vice versa. (Two setups open to the same aspirator are open to each other. If your system is at low pressure and the other is at atmospheric pressure, and they are suddenly joined, a dramatic and potentially destructive pressure surge will occur.)

To test your assembled apparatus for leaks, evacuate the system by closing stopcock B, opening stopcock A, and turning on the water full force. As the pressure in the apparatus decreases, the mercury levels in the arms of the manometer will move toward each other. After a while, the mercury levels will stop moving, indicating that you have reached the low-pressure limit of your aspirator. Close stopcock A firmly and watch the manometer closely. If the manometer levels move slowly apart, there is a leak in your system. If so, check your joints to be sure they are snug. You should not proceed with the experiment until your apparatus is airtight (or almost airtight. If there is a very slow leak, the experiment can still be performed.).

When the apparatus is leak-free, allow it to return to atmospheric pressure by opening stopcock B carefully. Disconnect the isoteniscope from the manifold and, holding the isoteniscope so that the open end is above the closed end, add your unknown organic liquid, using the capillary eyedropper you prepared in the glass-working session earlier.

WARNING: Some unknown organic vapors are poisonous or flammable. Do not leave your liquid in an open vessel for long periods of time. When disposing of organic liquid, dump it in the capped liquid waste disposal cans in the laboratory.

Fill until the level in the open end is about one and one-half inch below the level in the closed end (see Figure 4-3) when your isoteniscope is positioned in the beaker. Reconnect the isoteniscope to your apparatus.

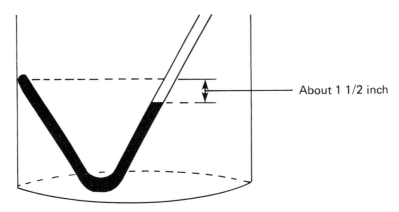

Figure 4-3. Isoteniscope in Beaker.

Now you must trap a small bubble of organic vapor in the closed end of the isoteniscope. (The presence of this bubble ensures that both a gas and a liquid phase will always be present so that the system can be at equilibrium). Proceed to form the bubble as follows. Introduce a bubble of air into the closed end by tilting the isoteniscope so that its plane is almost horizontal. Now by striking or flicking the closed end of the tube with your finger, break the bubble into many smaller ones. As these form, tilt the isoteniscope to allow all but one tiny bubble escape into the open end. Your goal is to isolate *one* very tiny bubble in the closed end.

Next, this bubble must be expanded and further fragmented. Close both stopcock A and stopcock B and turn on the aspirator. Carefully open stopcock A to lower the pressure in the system while you observe the small bubble. As soon as the bubble grows appreciably in size, close stopcock A and turn the aspirator off. Now fragment this bubble as before, again retaining only one small bubble. Warm the closed end of the isoteniscope with your fingers to cause the bubble to grow. Fragment as before. The small trapped bubble now consists almost entirely of organic vapor, and it provides the nucleus for the vapor phase which you will produce in this experiment.

Fill the 1000-ml beaker with cool water. Be sure the closed end of the isoteniscope is *entirely* submerged (why?). Lower the pressure in your apparatus slowly, keeping an eye on your trapped vapor bubble. Allow the bubble to grow until the liquid level in the closed arm of the isoteniscope is *below* the level in the open arm. (At this point, the vapor pressure of the trapped bubble is greater than the pressure indicated by the manometer.) Close stopcock A. Now, carefully open stopcock B very slightly and allow air to enter until the liquid levels in the two arms of the isoteniscope are identical. Close stopcock B at this point and read the column heights on your manometer. Record this

74

datum in your laboratory notebook. Record also the temperature of your water bath. If your aspirator is unable to lower the pressure enough to grow a bubble, warm your water bath by about 10°C and try again.

If your system has a slow leak which cannot be corrected, you should grow your vapor bubble as described above and close stopcock A. Then, instead of opening stopcock B, simply allow your slow leak to admit air to the system. One partner should watch the isoteniscope and, when he sees that the liquid levels are identical the other partner should read the manometer. (Note: It is possible to make a reading in just one arm of the manometer and still calculate the pressure. This requires only the assumption that the rise in one arm is equal to the fall in the other. However, if possible you should read both arms since the bore of your manometer may be somewhat irregular, vitiating the above assumption.)

After this first measurement, return your system about halfway to atmospheric pressure (enough so the bubble gets fairly small, but not so much that it disappears) and heat your water bath until it is about 10°C warmer. (If the bubble begins growing too large as you heat, bleed in some more air to compress it some more.) Now repeat the above procedure to measure the vapor pressure at this temperature.

The most likely cause of inadequate data in this experiment is failure to establish true temperature equilibrium at each point. The temperature of the bath will continue to rise after the burner is removed. You should anticipate this behavior and not record the temperature until it has stopped increasing. More important, be sure to allow enough time for the liquid in the isoteniscope to attain the temperature of the bath. Shaking the isoteniscope cautiously to agitate the liquid will hasten this process. Read the temperature at the same time that you establish the pressure balance.

Continue warming and taking data at about 10°C intervals until you reach a temperature where the vapor pressure exceeds atmospheric pressure (the bubbles will grow uncontrollably even though your apparatus is at atmospheric pressure) or until your water bath reaches ∿ 90°C, whichever occurs first.

To confirm the equilibrium (reversible) nature of your measurements, now take two additional vapor pressure values at lower temperatures between your last and first values. These will thus be within the range of the measurements already made, but will not duplicate them. By adding some cold water to the bath (and removing some of the water in it, if necessary) lower the temperature by 10 to 20 degrees. Adjust the system pressure and determine the vapor pressure of the sample. Again lower the temperature by another 10 to 20 degrees and repeat the vapor pressure measurement. (When plotting your data, use a different symbol for the values obtained during the cooling run, to distinguish them from the ones obtained on heating.)

Be sure to dispose of your unknown liquid in the capped liquid waste disposal can.

Treatment of Data

Refer to the section on "Making and Interpreting Line Graphs" in the introductory section (pages 25-32) as you proceed. Draw a graph of the vapor pressure of your unknown (y-axis) versus the temperature. Include this graph in your report.

The Clausius-Clapeyron equation

$$\log_{10} P = \frac{-\Delta H_v}{2.303\,R} \left(\frac{1}{T}\right) + I$$

is of the straight-line form $y = mx + b$ if $\log P$ is plotted on the y-axis and $1/T$ on the x-axis. Set up a table of $\log P$ versus $1/T$. Graph these points and draw the straight line through them which seems to come closest to all the points. The slope of this line, m, should be equal to $-\Delta H_v/2.303$ R, where $R = 8.3143$ J/deg mole. Evaluate ΔH_v from your graph. Calculate the units of ΔH_v (remember that the logarithm of a number is dimensionless). Using your calculated value of ΔH_v together with the T and P from a point *on your curve*, calculate I in the Clausius-Clapeyron equation.

Your instructor will have posted a list of organic compounds and their heats of vaporization. Use your experimental value of ΔH_v to try to pick out your compound from the list. Use your value of ΔH_v and the Clausius-Clapeyron equation to calculate the boiling point of your compound. Also, get a value for the boiling point by graphical extrapolation of your $\log P - T^{-1}$ plot. Compare these with the tabulated value.

Be sure to organize your notebook-report logically and neatly. It is very important to take care with significant figures in this experiment.

In your error analysis section, try to estimate the precision in pressure and temperature for each point. Do all your points fall on a straight line to within these limits when $\log P$ is plotted against $1/T$? Given your plotted points, what are the extremes for reasonable straight lines you can fit them to? What range of possible ΔH_v does this give? Does your value for ΔH_v agree with a literature value to within these error limits?

Answer the following questions in your report.

a.   Using identical apparatus, two men begin to heat equal amounts of water, initially at $25°C$. One man is working at sea level, the other on a high mountaintop. Which man will observe boiling first? Explain.

b.   Each man in question (a) had an uncooked egg in his water when he started heating. Which egg gets hard-boiled first? Explain.

c.   What advantage comes from using a pressure cooker? How does it work?

d.   Suppose your unknown liquid contained a significant amount of dissolved air. Would you expect this to affect your measured value of ΔH_v? If so, how would it be affected?

Questions

Be sure you can answer these when you come to lab. There may be a quiz.

1.   What hazards may be associated with your unknown organic liquid? How will you dispose of it?
2.   Why are pressure readings taken only when the liquid levels in the two arms of the isteniscope are identical?
3.   What is the purpose of the ballast bottle?
4.   What is the purpose of the trap bottle?
5.   How are the column heights in the manometer related to pressure?
6.   In what way is mercury hazardous?
7.   What is the function of stopcock A? Of stopcock B?
8.   What is the purpose of the sequence of growing and fragmenting operations on the isteniscope bubble?
9.   Why are you cautioned (p. 73) to close stopcock A before turning off the water to the aspirator?

Chemical Kinetics

The purpose of this experiment is to measure the effects of changes in concentration and temperature on the rate of a reaction.

Theory

Definition of Reaction Rate. Suppose a solution is prepared to be initially 1 M in substances A and B. Suppose also that the substances A and B react to form C and D:

$$A + B \rightarrow C + D. \tag{1}$$

After 50 seconds, say, the concentrations of C and D have built up to 0.1 M while the concentrations of A and B have dropped to 0.9 M. The **average rate of the reaction** is expressed as the amount of reagent (in units of molarity) which is used up or produced per unit time. Thus, in this example

$$\text{Rate} = -\Delta[A]/\Delta t \equiv \frac{(\text{minus the change in concentration of A})}{(\text{the change in time})} =$$

$$\frac{0.1 \text{ moles/liter}}{50 \text{ seconds}} = 0.002 \text{ moles/liter seconds}.$$

The same rate would result for $-\Delta[B]/\Delta t$, $\Delta[C]/\Delta t$, or $\Delta[D]/\Delta t$. (The negative sign is used to compensate for the fact that, since A and B are disappearing, $\Delta[A]$ and $\Delta[B]$ are negative quantities.) It is often important to specify which reagent is being referred to by a rate. If the stoichiometry were $2A + B \rightarrow C + 3D$, the rate of disappearance of A would be twice that for B, and the rate of appearance of D would be greater still, so care must be taken in defining what is meant by the "reaction rate."

In the above discussion, we simply took the concentration of a reagent at two different times and calculated an average rate. But certain complications may prevent this from being the correct rate for reaction 1. Suppose we are measuring the rate of disappearance of A. Then we must be sure that all the A which disappears does so according to equation 1 (i.e., that there are no "side reactions" using up A)—the reaction must be clean. We must also be sure that no A is being produced by the reverse reaction:

$$C + D \rightarrow A + B. \tag{2}$$

—the reaction must be **irreversible**. One way to guarantee that a reaction be essentially irreversible is to remove a product (C or D) from the solution as it is formed. If a product is a precipitate or an insoluble gas, this happens automatically. Otherwise, it may be necessary to add something which reacts with C or D in such a way as to remove it. Alternatively, one could measure the **initial reaction rate** at the very beginning of the reaction, before the product concentration has built up to the point where significant back reaction can occur.

Effects on Reaction Rate of Changing Concentration. In order for two molecules of reactant to react, it is necessary for them to come together in solution (i.e. to collide) with sufficient energy to rearrange and form products. Thus, of the collisions occurring between molecules A and B, many will be unfruitful due to inadequate energy. Suppose that, at room temperature, ten percent of the collisions between A and B are fruitful. What will be the effect of doubling the concentration of B? Obviously each A molecule will then encounter a B molecule twice as often, so the rate of collision will double. Of these collisions, ten percent will still be fruitful, so the reaction rate should double. Exactly the same reasoning applies for the effect of doubling the concentration of A. Thus, *assuming that the reaction rate is controlled by collisions between A and B*, the effect of doubling the concentration of A or B should be to double the rate of reaction. This can be mathematically summarized by a **rate law**:

$$\text{Rate} \equiv \frac{-\Delta[A]}{\Delta t} = k[A][B]. \tag{3}$$

Here the square brackets symbolize units of molarity, and k, the **rate constant**, is the proportionality constant relating the observed rate to reagent concentration. We see that the rate law (3) states that the rate will double if [A] or [B] is doubled. Because the concentrations of A and B appear in (3) raised to the first power, this rate expression is said to be **first order** in A and first order in B.

In general, it is *not* possible to obtain the rate law by inspecting the balanced equation for a reaction. This is because the overall reaction may be the net result of a sequence of subreactions. If one of these subreactions is particularly slow, it will act as a "bottleneck" for the process, thereby determining the reaction rate. Thus, it is the collision involved in this unknown **rate determining step** that is reflected in the rate law. The rate law, then, can be determined only be recourse to experiment. The rate law often has the form

$$\text{Rate} = k[A]^a[B]^b[C]^c \ldots \tag{4}$$

where A, B, C, . . . are all reactants for the reaction being studied, and a, b, c, . . . are numbers obtained from experiment. These numbers may be positive, zero, or negative, and do not have to be integers. Thus, suppose you found that doubling the concentration of A caused a reaction to go four times as fast, that doubling the concentration of B had no effect, and that doubling the concentration of C increased the rate by a factor of 1.4. The reaction would be second order in A, zero order in B, and one-half order in C:

$$\text{Rate} = k[A]^2[B]^0[C]^{1/2} = k[A]^2[C]^{1/2}.$$

[Recall that $2^{1/2} = \sqrt{2} = 1.414$, and $x^0 = 1$ for any number x.]

Effect on Reaction Rate of Temperature Change. We have already mentioned that a certain amount of energy of motion must be present for a collision between reactant molecules to be fruitful. The forces holding the nuclei together in reactant molecules must be overcome during the process of formation of product molecules. This process is depicted schematically by the diagram of potential energy versus the progress of a reaction in Figure 5-1.

This diagram shows that energy will be released if A and B are transformed to C + D. But an amount of energy, E_b (the **barrier energy**), is needed to deform A and B sufficiently so that the bonds in C and D can begin to form. A collision will be fruitful only if A and B collide with a kinetic energy large enough to traverse the potential energy barrier of height E_b.

Since increasing the temperature of a reaction mixture increases molecular velocities, it is reasonable that reaction rates increase also, since the percentage of fruitful collisions increases. (Another factor also operates: The *frequency* of intermolecular collisions increases. But this turns out to be a minor effect, and we shall ignore it.) That the reaction rate increases with heating must be reflected by an increased value of the rate constant, k. The Arrhenius equation relates k to the temperature:

$$k = Ze^{-E^*/RT} \tag{5}$$

78

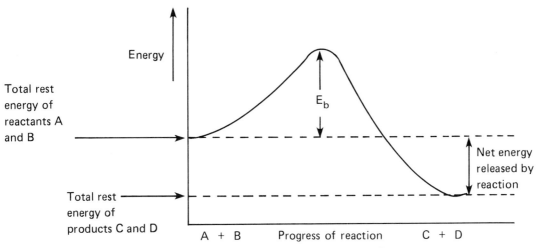

Figure 5-1. Energy Profile for a General Reaction.

where Z is the frequency factor, e is the base of the natural logarithms, R is the gas constant (8.3143 joules/deg mole), T is the absolute temperature, and E^* is the **activation energy**. It has approximately the same value as the barrier height, E_b.

For further discussion of chemical kinetics, see:

Brown and LeMay	Ch. 13 (375-389)
Masterson and Slowinski	Ch. 16 (377-390)
Mortimer	Ch. 13 (435-442)
Sienko and Plane	Ch. 10 (280-288)

Description of the Iodine Clock Reaction. Your task is to determine by experiment the orders and activation energy, E^*, for a particular reaction. The reaction you are concerned with is the oxidation of iodide by hydrogen peroxide:

$$H_2O_2 + 2\,I^- + 2\,H^+ \rightarrow I_2 + 2\,H_2O. \qquad (6)$$

Also present in the solution will be a known amount of thiosulfate ion, $S_2O_3^{2-}$, and some dissolved starch. The thiosulfate is present in order to reduce the iodine product back to iodide reagent just as fast as it is formed:

$$2\,S_2O_3^{2-} + I_2 \rightarrow 2\,I^- + S_4O_6^{2-} \qquad (7)$$

The dissolved starch is an indicator for iodine, forming a deep blue color when the I_2 concentration begins to build up. The following is what happens, then:

1. A solution is prepared containing known concentrations of I^-, $S_2O_3^{2-}$, H^+, and some starch.
2. A known amount of H_2O_2 is stirred in rapidly and the time is noted.
3. The H_2O_2 oxidizes I^- to I_2, but the I_2 is immediately reduced back to I^- by the thiosulfate. The solution remains water clear.
4. After a while the thiosulfate gets all used up. Then the I_2 is no longer removed, so the solution turns blue. This time is noted.

79

Thus you will have measured an elapsed time, Δt. You will also know how much thiosulfate was used up (all that was present). The balanced equation 7 then tells you how much I_2 was reduced. But the I_2 reduced in equation 7 is identical in amount to the I_2 produced in equation 6. And equation 6 tells us that the amount of I_2 produced is equal to the amount of H_2O_2 used up. So you will have indirectly measured $-\Delta[H_2O_2]$. This, divided by Δt, gives the rate of the reaction.

Note that the rate of the reaction between H_2O_2 and I^- is not influenced by the presence of thiosulfate ion, $S_2O_3^{2-}$. So long as thiosulfate is present, it reacts instantaneously with the I_2. Thus, during the reaction (until the blue color develops) the thiosulfate ion concentration decreases at a rate determined by the H_2O_2–I^- reaction rate, which in turn is fixed by the concentrations of H_2O_2 and I^-.

You will not measure the dependence of the rate on $[H^+]$. A buffer solution will be present in your reaction mixtures and this will ensure that $[H^+]$ remains constant. Notice that there is no problem with the reverse reaction in this experiment. This is because one of the products, I_2 is removed by the thiosulfate as fast as it is produced, thereby preventing the reverse of equation (6). The reaction is also clean, provided that you follow the precautions regarding cleanliness in the EXPERIMENTAL section below.

In calculating the reaction orders and rate constant, you will be making an *assumption*—namely that the initial concentrations of I^- and H_2O_2 pertain throughout the entirety of the measured reaction. Will this assumption produce a serious error? With I^- there is no problem. It is transformed back into I^- just as fast as it is oxidized to I_2, so its concentration is indeed constant. This is *not* true for H_2O_2. It gets used up and is not regenerated. However, by always adding to our starting solution at least ten times as much H_2O_2 as we need, we guarantee that the change in concentration of H_2O_2 will be small enough that we can safely use the initial value.

You will perform this reaction six times with solutions of known concentration and twice using a solution of unknown concentration. Some of these runs will differ from each other only in $[H_2O_2]$, others only in $[I^-]$, and some only in temperature. By mathematically comparing these rates, you will be able to obtain the orders of the reaction in H_2O_2 and in I^-, and also the activation energy. Detailed directions for doing this follow the EXPERIMENTAL section.

Experimental Procedure

Two students are to work together on this experiment.
Submit separate reports and analyze separate unknowns.

NOTES:

1. This experiment strains distilled water resources. Be extra careful not to waste it.
2. Hydrogen peroxide reacts with many common substances. Hence your equipment should be *scrupulously clean* for this experiment.
3. For washing your glassware use detergent and a brush. Rinse thoroughly with tap water. Then finally rinse with small portions of distilled water. Shake out excess water but do not dry.
4. The sodium thiosulfate solution and the hydrogen peroxide solution are nominally 0.02 M and 0.1 M, respectively. The exact concentrations, indicated on the stock bottles, will vary from section to section.

Secure the indicated amounts of the following solutions from the side shelf stock bottles. Since the KI, $Na_2S_2O_3$, and H_2O_2 solutions are of exactly known concentrations, the containers used for them must be rinsed with small portions of solution before you add the supply you are going to use. For each solution there is a graduated cylinder to be used* in measuring out the amount to be transferred to your clean labelled container.

a. 210 ml buffer solution (0.5 M acetic acid, 0.5 M sodium acetate). Take this solution in a 250-ml beaker.
b. 135 ml 0.300 M potassium iodide solution. Take this solution in a 250-ml beaker.
c. 18 ml starch solution. Take this in a 50-ml beaker.
d. 210 ml ca. 0.02 M sodium thiosulfate solution. (See the label on the dispensing bottle for exact molarity.) Take this solution in a 250-ml beaker.
e. 350 ml ca. 0.1 M hydrogen peroxide. (See the label on the dispensing bottle for exact molarity.) Take this solution in a 400-ml beaker.

First, perform the following qualitative tests before you do the kinetic runs. In a test tube add a few drops of H_2O_2 solution to approximately 5 ml of KI soltuion. Record your observations and interpretation. Divide the mixture into three portions and use one as a standard. To a second portion, add a few drops of starch solution. Again, record observations and interpretation. Finally, to the third add a few milliliters of thiosulfate solution and then a few drops of starch, making a record of your observations in each step. Discuss the relation of these tests to the chemistry occurring during the kinetic runs.

You will now make six kinetic runs using solutions made up according to those in the chart below. Each solution has a total volume of 500 ml.

	Distilled Water	Buffer	$Na_2S_2O_3$ ($\sim$0.02 M)	Starch	KI (0.300 M)	H_2O_2 ($\sim$0.1 M) [Do not add until ready to time reaction]
A	415 ml	25 ml	25 ml	2 ml	8 ml	25 ml
B	408 ml	25 ml	25 ml	2 ml	15 ml	25 ml
C	398 ml	25 ml	25 ml	2 ml	25 ml	25 ml
D	373 ml	25 ml	25 ml	2 ml	25 ml	50 ml
E	323 ml	25 ml	25 ml	2 ml	25 ml	100 ml
F	373 ml	25 ml	25 ml	2 ml	25 ml	50 ml

In a clean 600-ml beaker mix the components of solution A in the order indicated **except** for the hydrogen peroxide. Use your burets for measuring the KI and the $Na_2S_2O_3$ solutions. (Remember the usual procedure to avoid changing the concentration of a solution when it is added to a wet vessel.) Place a sheet of white paper under the beaker to aid in detecting the time of the first color change. One student should record the time to the nearest second as the other student adds the H_2O_2. (It is suggested you perform this step as the second hand sweeps across "12" on your watch.) *While stirring the solution*, the second student should pour the 25 ml of H_2O_2 in rapidly (use a 25-ml graduated cylinder). Continue to stir the solution thoroughly for another 10-15 seconds. Note the temperature

*If your beakers have volume graduations, you may measure the solutions directly without using the cylinders since only an approximate volume measurement is needed. Do not take an excess of solution, however.

of the solution. [It will be necessary for one student to keep an eye on this solution so as not to miss the precise second that the color changes. However, this may take quite awhile (times seem to vary between 10 and 30 minutes depending on the purity of the distilled water). Hence it is advisable that you proceed with other runs while waiting for solution A to finish.]

When the solution changes color, record the time to the nearest second in your laboratory notebooks (both partners). Discard the solution; clean the beaker and graduated cylinder by thoroughly rinsing, first with tap water, then twice with small amounts of distilled water. Shake out excess water. Do *not* dry them with a towel. (That would be unscrupulous!)

Make the runs using the next four solutions, B-E, in a similar manner. It is important that the temperatures of all the solutions A-E be the same. If necessary, adjust the temperature of each succeeding solution to within 0.5°C of that of the first solution before adding the peroxide. The temperature should be around room temperature (18-25°C). Note that solution E requires you to use your 100-ml graduated cylinder for quick addition of H_2O_2.

For run F prepare an ice-water mixture in your 1000-ml beaker. Mix together the chemicals for solution F in the usual way (omitting the peroxide) in your 600-ml beaker. Now place the beaker containing solution F into the ice-water bath and stir until the solution is about 10°C colder than were the solutions A-E. Separately cool the 50 ml of H_2O_2 to the same temperature. (Avoid transferring a thermometer from one solution to the other or you will contaminate them.) When all is ready, remove the solutions from the cooling bath, mix in the cooled H_2O_2 as before, and record the time necessary for a color change to occur. It will be necessary for you to stir the solution to keep the temperature uniform. You will also need to occasionally replace the beaker into the bath to keep the temperature from rising. Keep the temperature as constant as you can for the duration of the run.

Each partner should submit a clean, labelled 100-ml volumetric flask to his instructor. This will be returned with some KI solution. Dilute it to the mark with distilled water. See page 36 for specific instructions; your wash bottle is useful in making the final addition of the last few mls. Using 25 ml of his unknown solution, each partner should prepare and time one more run, made up according to the directions for solution C. The observed rate for this run will be used to estimate the concentration of the KI solution.

Treatment of Data

Determining the Order of the Reaction. Since the buffer solution prevents variation of [H⁺], this concentration dependence is suppressed and gets incorporated into the rate constant k. The rate expression, then, is

$$\text{rate} = \frac{-\Delta [H_2O_2]}{\Delta t} = k\,[H_2O_2]^r[I^-]^s. \tag{8}$$

Taking the logarithm of both sides gives us either

$$\log(\text{rate}) = \log(k[H_2O_2]^r) + s \log[I^-] \tag{9a}$$

or

$$\log(\text{rate}) = r \log[H_2O_2] + \log(k[I^-]^s). \tag{9b}$$

Other separations are possible, but these are the ones we need.

Determining s, the Order in I⁻. Note that in solutions A, B, and C, the concentrations of all chemicals are identical except for iodide ion. Thus, for these three runs, the first term on the right hand side of equation 9a is a constant, whereas log (rate) and log [I⁻] vary. Hence equation

9a is a linear equation of the familiar form $y = b + mx$. By plotting log (rate) on the ordinate (y-axis) against log $[I^-]$ on the abscissa for runs A-C, you should obtain a straight line having a slope of s. Round off your value of s to the nearest integer or half-integer. Attach your graph to your report. [Recall the details of preparing a graph and determining the slope. See pages 26-30.]

Determining r, the Order in H_2O_2. Solutions C, D, and E differ only in the concentration of H_2O_2. Utilizing equation 9b gives a linear equation when log (rate) is plotted against log $[H_2O_2]$. The slope, rounded off as before, is your value for r. Attach this graph to your report.

Calculating the Rate Constant, k. Using your values for r and s, and your known values for rate, $[I^-]$, and $[H_2O_2]$ for each run A-E, calculate a value for k for each run using equation 8. Calculate from these an average value of k. Analyze the units in equation 8 to figure out the units for k. Notice that the units of k depend on your values for r and s.

Calculating the Activation Energy, E*. We have already mentioned that the rate constant is related to temperature by the formula

$$k = Ze^{-E*/RT} \tag{10}$$

Taking the natural logarithm of both sides of (10) gives

$$\ln(k) = \ln(Z) - E*/RT. \tag{11}$$

At another temperature, T', we can expect some other rate constant, k', which will be given by

$$\ln(k') = \ln(Z) - E*/RT'. \tag{12}$$

Subtracting (11) from (12) and rearranging gives

$$\ln\left(\frac{k'}{k}\right) = \frac{E*}{R}\left(\frac{T' - T}{TT'}\right) \tag{13}$$

or, in base 10 logarithms (after further rearranging)

$$E* = \frac{2.303\,RTT'}{(T' - T)}\,\log\left(\frac{k'}{k}\right). \tag{14}$$

(R is equal to 8.3143 Joules/deg mole). Thus, to calculate $E*$, you need to know the ratio of rate constants at two temperatures.

You have measured the reaction times for two solutions (D and F) which are identical except for temperature. This means that any difference in reaction times must be due to changes in k. From your relative times, then, you can obtain the rate constant ratio, k'/k. But note that this is a reciprocal relation: If the rate constant *doubles,* the time for reaction is halved. Therefore, $\frac{\Delta t}{\Delta t'} = \frac{k'}{k}$. If you rewrite equation 14 with Δt's and use it for calculating $E*$, you avoid getting a wrong value in case you have made any error in calculating the k's. Calculate $E*$ and report your value. Analyze the units in equation 14 to determine the units for $E*$, and include this in your report.

Determining the Concentration of Your Unknown. Use your graph of log (rate) vs. log $[I^-]$ for runs A, B, and C to estimate the concentration of KI in your 100-ml volumetric flask. [The practice of using *rate* measurements to indirectly determine *concentration* is common in biochemical studies.]

Organization of the Notebook and Error Discussion

Since you will be making a series of related runs, it will pay to have tables set up to receive data and to summarize results of calculations. To begin with, you will be measuring initial time, final time, and temperature for seven solutions (A-F and your unknown). Have a table ready for these data. From these data and the recipes for making the solutions, you will be calculating $[I^-]$, $\log [I^-]$, $[H_2O_2]$, $\log [H_2O_2]$, elapsed time in seconds (Δt), rate ($= -\Delta[H_2O_2]/\Delta t$), and log (rate). Set up a table to display these results. You need show but one sample calculation for each type of result. Be careful with significant figures.

In calculating $-\Delta[H_2O_2]$, you should calculate

a. moles $S_2O_3{}^{2-}$ in the solution.
b. number of moles of I_2 required to exhaust the moles of $S_2O_3{}^{2-}$ from (a).
c. number of moles of H_2O_2 required to produce the number of moles of I_2 in (b).
d. the change in *concentration*, $-\Delta[H_2O_2]$, corresponding to loss of number of moles of H_2O_2 in (c). Recall that the solution volume is 500 ml.

Note that you need perform this series of calculations only once since the same amount of $S_2O_3{}^{2-}$ is used in every run.

In your error discussion, present your analysis of the uncertainty, or error limits, for the rates. Assume that the stock solution concentrations are correct to the number of significant figures indicated on the labels. State what you believe to be the chief sources of the errors in rate. What are the extremes of slope you could reasonably achieve in fitting your plotted data points? Would these extremes embrace any values of r and s different from those you selected? How much error range would you give to your estimate of KI concentration?

Questions

Be sure you can answer these when you come to lab. There may be a quiz.

1. A reaction proceeds according to the balanced equation

$$2 A + B \rightarrow 2 C.$$

   a. Write down three expressions for the rate of this reaction. How would the rate of disappearance of B compare to the rate of appearance of C?
   b. The reaction is found to proceed nine times as fast if the concentration of A is tripled, and three times as fast if the concentration of B is tripled. Write down the rate law. What is the order of the reaction in A? In B?

2. What is the balanced equation for the reaction you are studying? What is the equation for the subsequent reduction of iodine?
3. What is the purpose of the starch solution?
4. If 0.00025 moles of $Na_2S_2O_3$ is present in the starting solution, how many moles of H_2O_2 are used up by the time the solution turns blue? (See equations 6 and 7.) If the total volume of the solution is 500 ml, what is $-\Delta[H_2O_2]$? If this reaction requires 600 seconds, what is the value of the rate? Show units.
5. What is the purpose of the buffer solution?
6. What would be observed when the H_2O_2 was added in a kinetic run if, by mistake, the $Na_2S_2O_3$ solution had been left out?
7. Why is cleanliness of great importance in this experiment?

Appendix

The Vapor Pressure of Water at Different Temperatures

Temperature (°C)	Vapor Pressure (mm of mercury)	Temperature (°C)	Vapor Pressure (mm of mercury)	Temperature (°C)	Vapor Pressure (mm of mercury)
−10 (ice)	1.0	21	18.6	40	55.3
− 5 (ice)	3.0	22	19.8	45	71.9
0	4.6	23	21.1	50	92.5
5	6.5	24	22.4	60	149.4
10	9.2	25	23.8	70	233.7
15	12.8	26	25.2	80	355.1
16	13.6	27	26.7	90	525.8
17	14.5	28	28.3	100	760.0
18	15.5	29	30.0	110	1074.6
19	16.5	30	31.8	150	3570.5
20	17.5	35	42.2	200	11659.2

Composition of Commerical "Concentrated" Reagents

		For Wt.	% Solute*	Density*	Molarity*
Hydrochloric acid	HCl	36.46	37.4	1.19 g/cm^3	12.2
Sulfuric acid	H_2SO_4	98.08	96.0	1.84	18.0
Nitric acid	HNO_3	63.01	70	1.42	15.8
Acetic acid	$HC_2H_3O_2$	60.05	99.7[+]	1.05	17.4
Aqueous ammonia	NH_3	35.05	58	0.90	14.9

*Approximate values; vary slightly with supplier.

Special (non-desk) equipment

Exp. 1   Calorimeter (styrofoam cups)

Exp. 2   Mohr buret
Leveling bulb with connecting tubing
Rubber nipple for buret

Exp. 4   Mercury manometer with vacuum manifold
Vacuum tubing

FISHER SCIENTIFIC / PERIODIC CHART OF THE ELEMENTS

IA	IIA	IIIB	IVB	VB	VIB	VIIB	VIII			IB	IIB	IIIA	IVA	VA	VIA	VIIA	NOBLE GASES
1 H 1.0079																	2 He 4.00260
3 Li 6.941	4 Be 9.01218											5 B 10.81	6 C 12.011	7 N 14.0067	8 O 15.9994	9 F 18.99840	10 Ne 20.179
11 Na 22.98977	12 Mg 24.305											13 Al 26.98154	14 Si 28.086	15 P 30.97376	16 S 32.06	17 Cl 35.453	18 Ar 39.948
19 K 39.098	20 Ca 40.08	21 Sc 44.9559	22 Ti 47.90	23 V 50.9414	24 Cr 51.996	25 Mn 54.9380	26 Fe 55.847	27 Co 58.9332	28 Ni 58.71	29 Cu 63.546	30 Zn 65.38	31 Ga 69.72	32 Ge 72.59	33 As 74.9216	34 Se 78.96	35 Br 79.904	36 Kr 83.80
37 Rb 85.4678	38 Sr 87.62	39 Y 88.9059	40 Zr 91.22	41 Nb 92.9064	42 Mo 95.94	43 Tc 98.9062	44 Ru 101.07	45 Rh 102.9055	46 Pd 106.4	47 Ag 107.868	48 Cd 112.40	49 In 114.82	50 Sn 118.69	51 Sb 121.75	52 Te 127.60	53 I 126.9045	54 Xe 131.30
55 Cs 132.9054	56 Ba 137.34	57 *La 138.9055	72 Hf 178.49	73 Ta 180.9479	74 W 183.85	75 Re 186.2	76 Os 190.2	77 Ir 192.22	78 Pt 195.09	79 Au 196.9665	80 Hg 200.59	81 Tl 204.37	82 Pb 207.2	83 Bi 208.9804	84 Po (210)	85 At (210)	86 Rn (222)
87 Fr (223)	88 Ra 226.0254	89 ▼Ac (227)	104 § (260)	105 § (260)													

★ Lanthanoid Series

58 Ce 140.12	59 Pr 140.9077	60 Nd 144.24	61 Pm (147)	62 Sm 150.4	63 Eu 151.96	64 Gd 157.25	65 Tb 158.9254	66 Dy 162.50	67 Ho 164.9304	68 Er 167.26	69 Tm 168.9342	70 Yb 173.04	71 Lu 174.97

† Actinoid Series

90 Th 232.0381	91 Pa 231.0359	92 U 238.029	93 Np 237.0482	94 Pu (244)	95 Am (243)	96 Cm (247)	97 Bk (247)	98 Cf (251)	99 Es (254)	100 Fm (257)	101 Md (258)	102 No (255)	103 Lr (256)

FISHER SCIENTIFIC COMPANY
CAT NO. 5-702-5

§ The International Union for Pure and Applied Chemistry has not adopted official names or symbols for these elements

† These weights are considered reliable to 3 in the last place. Other weights are reliable to 1 in the last place.

Atomic weights corrected to conform to the 1971 values of the Commission on Atomic Weights.

Copyright 1971 by Fisher Scientific Company

87

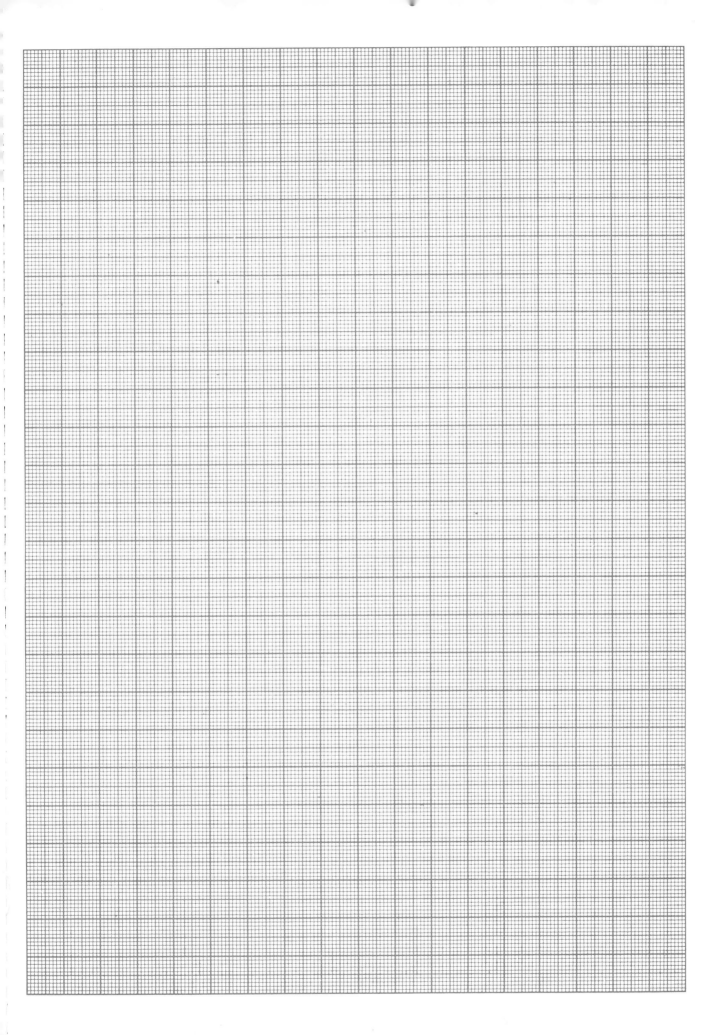

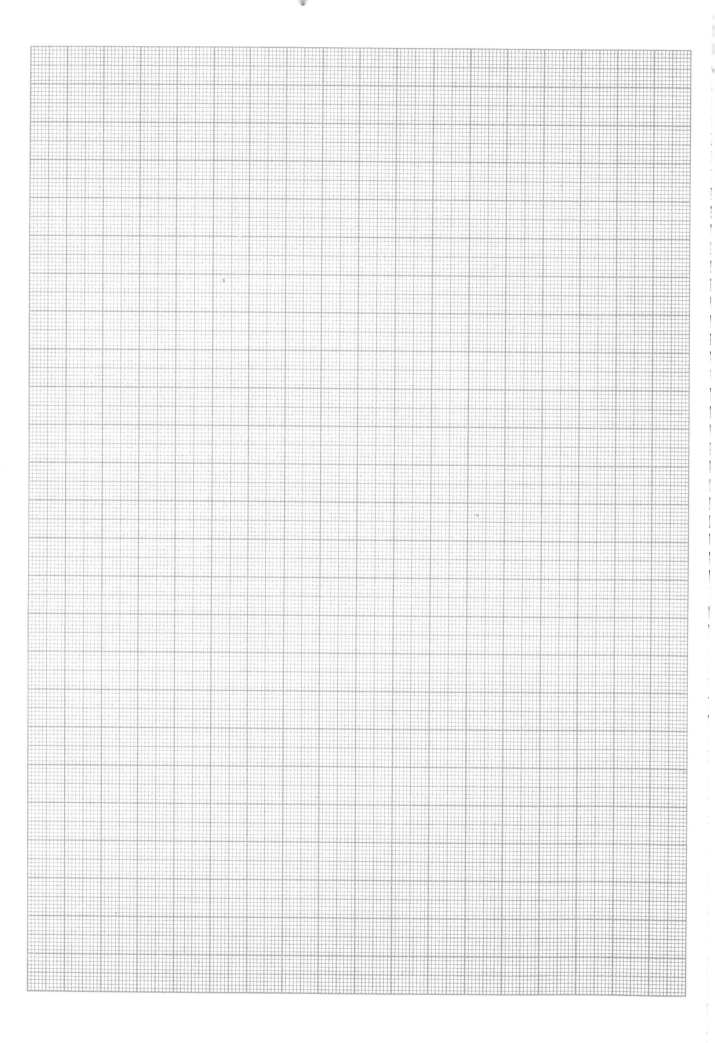

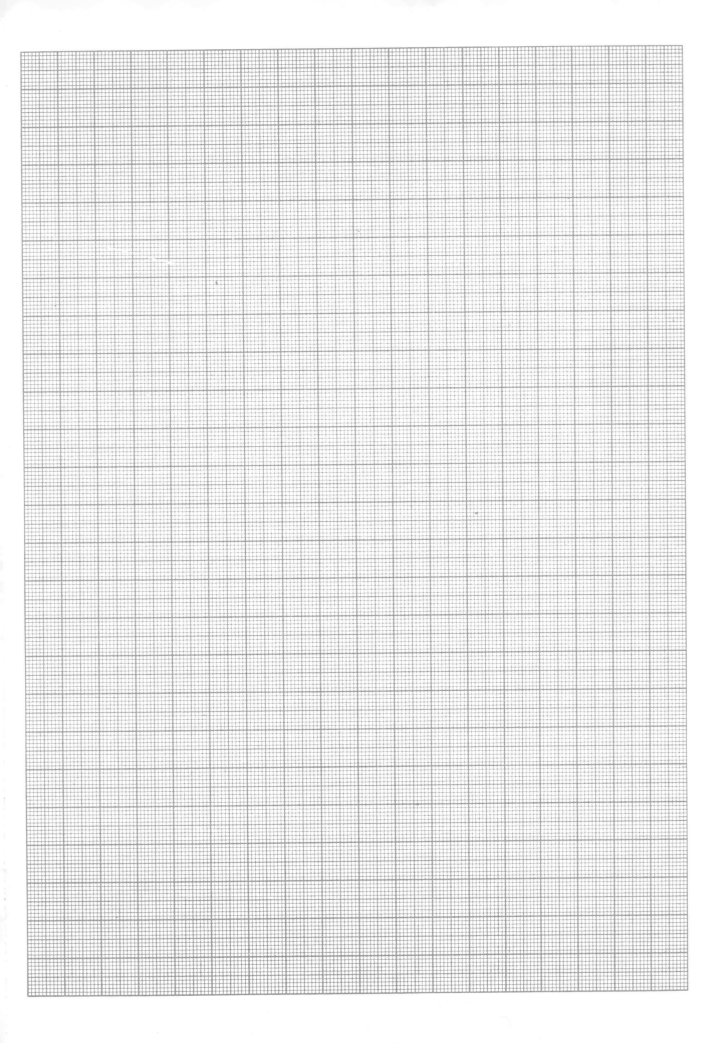

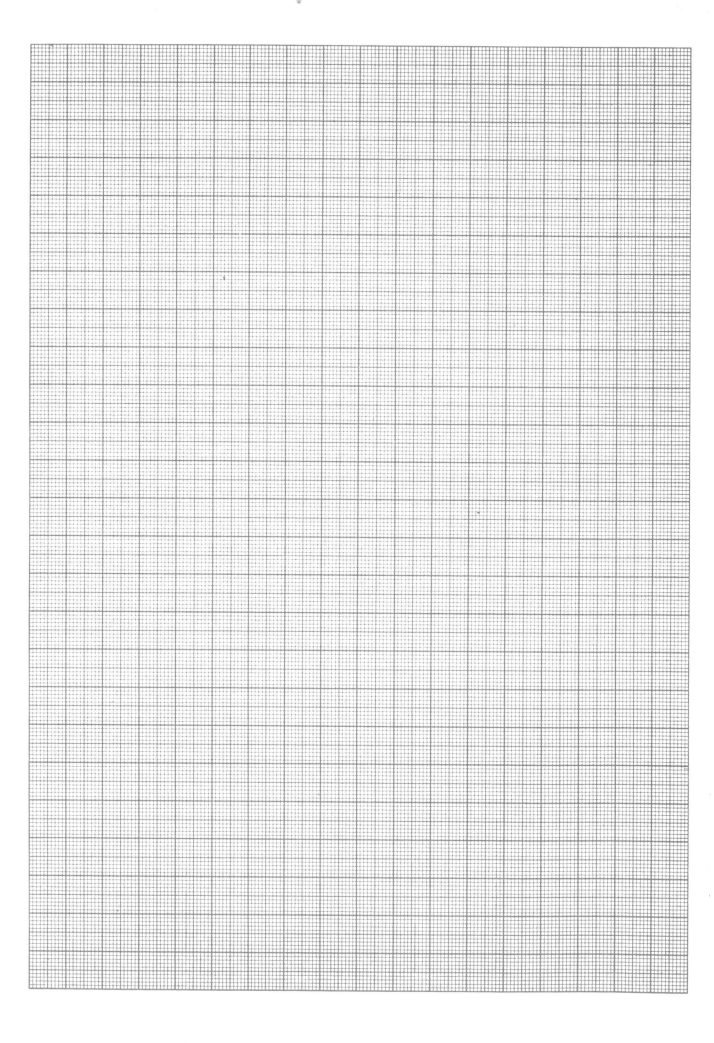

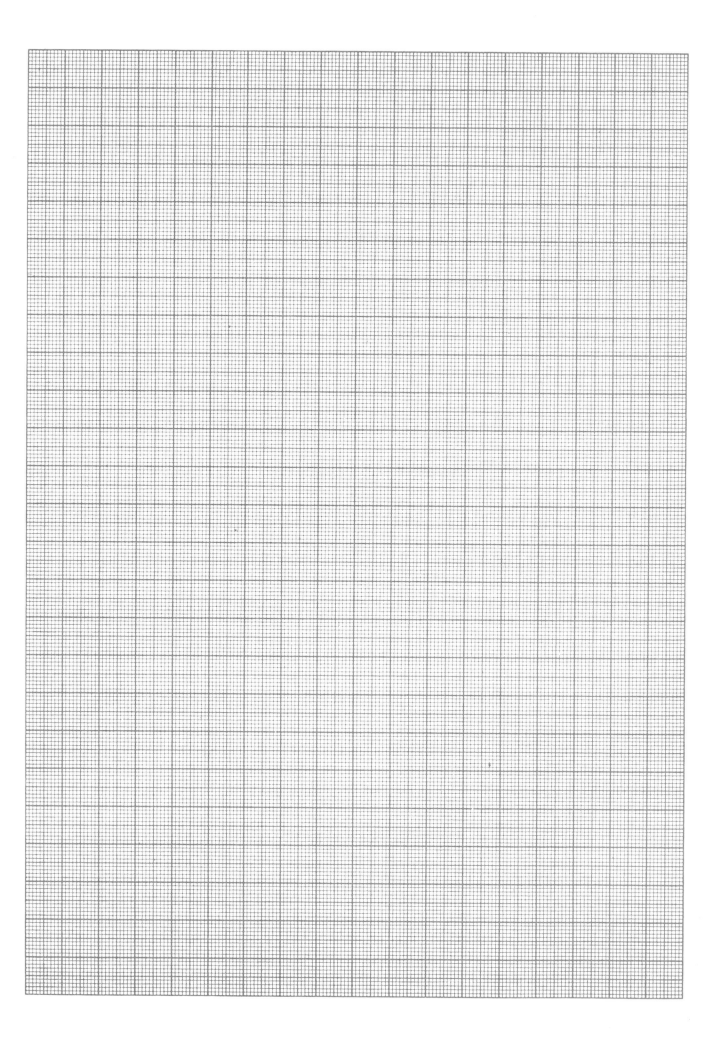

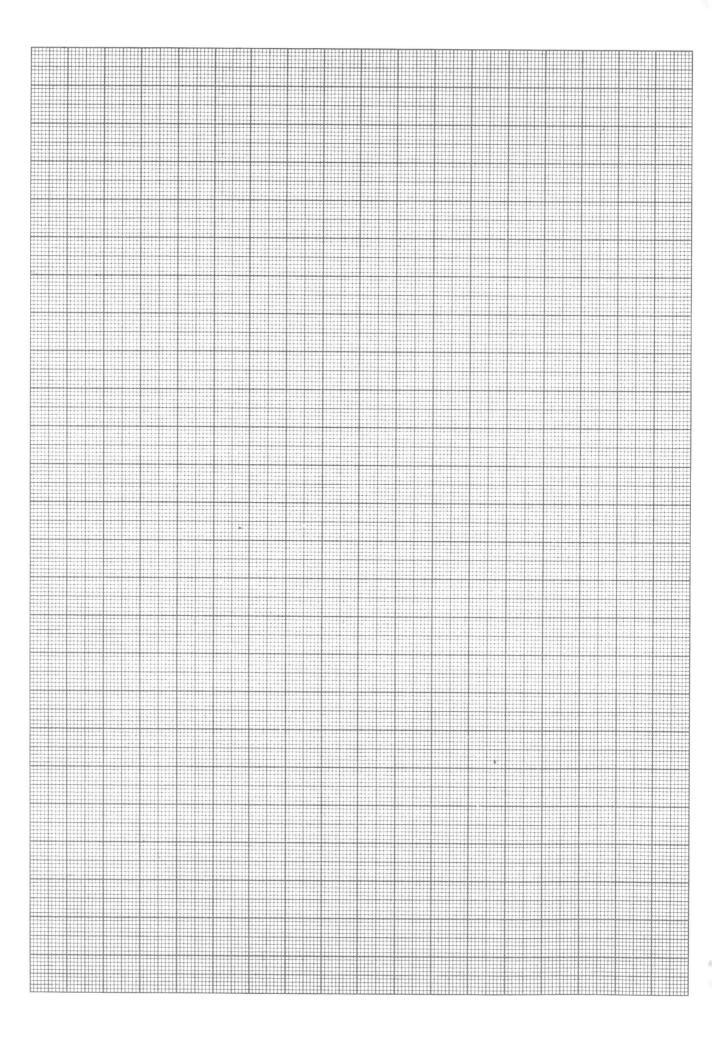

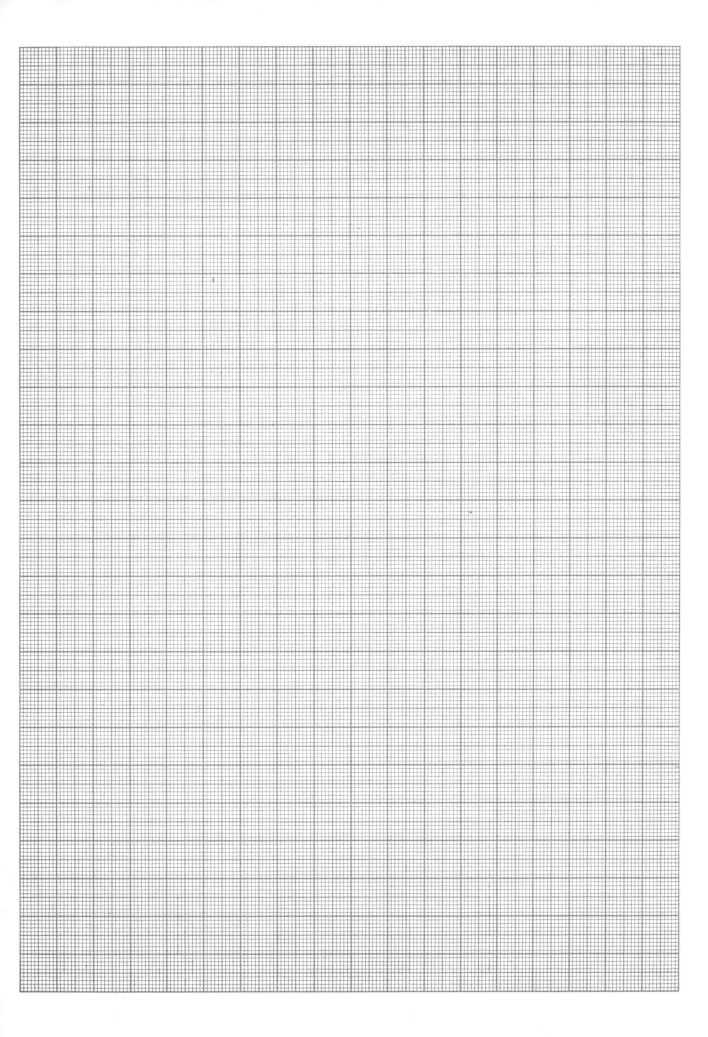

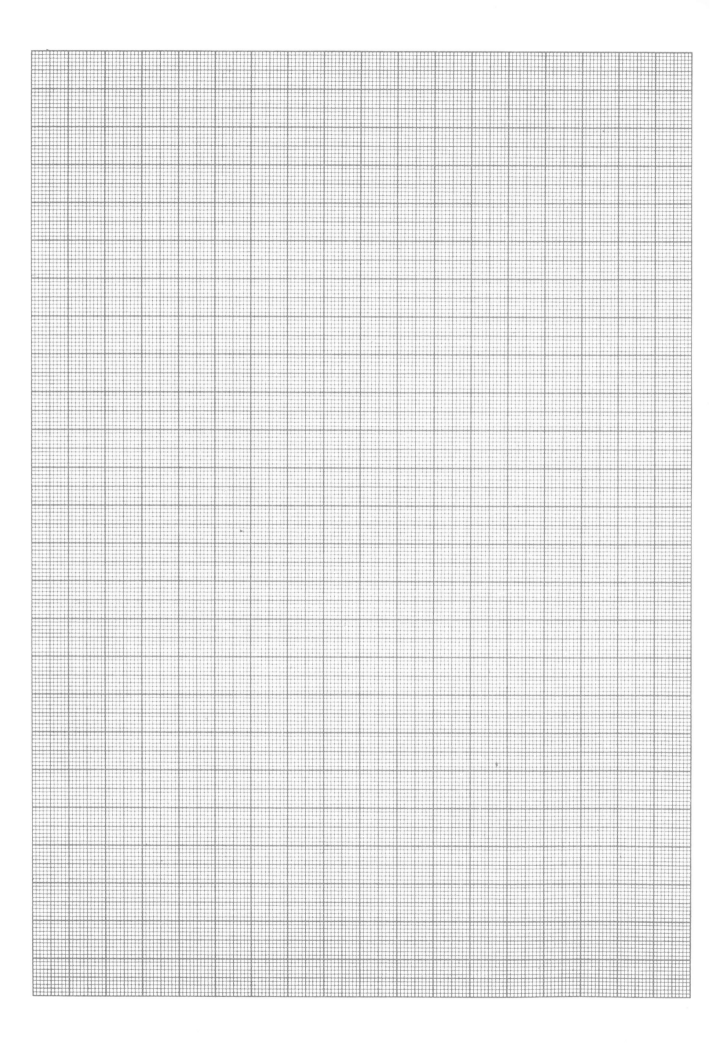

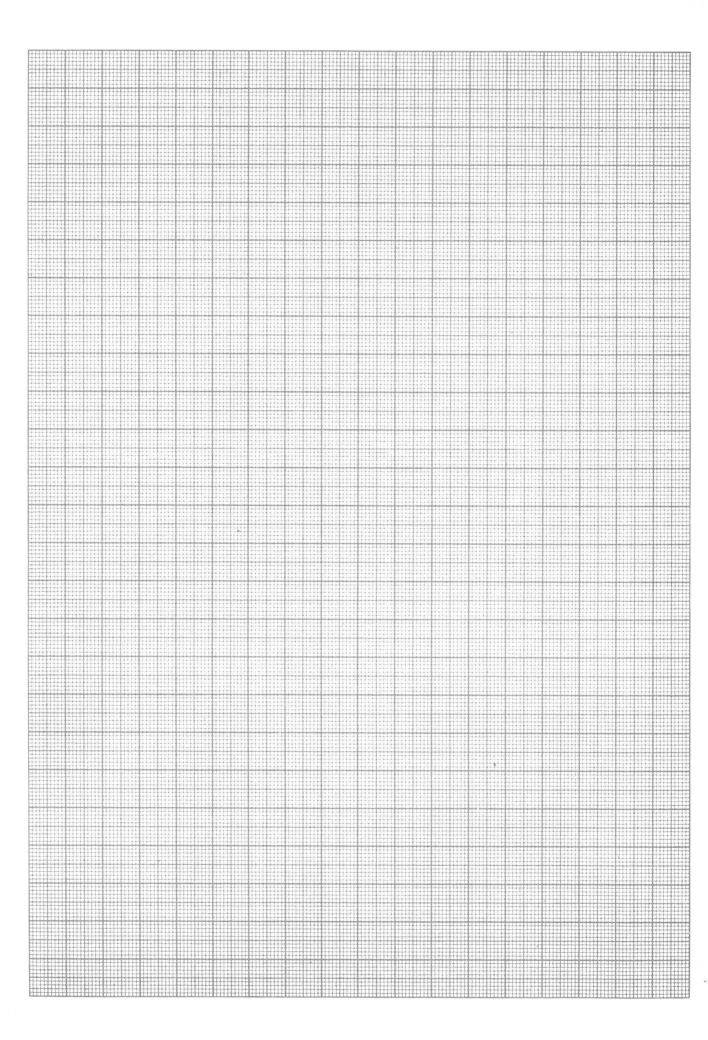

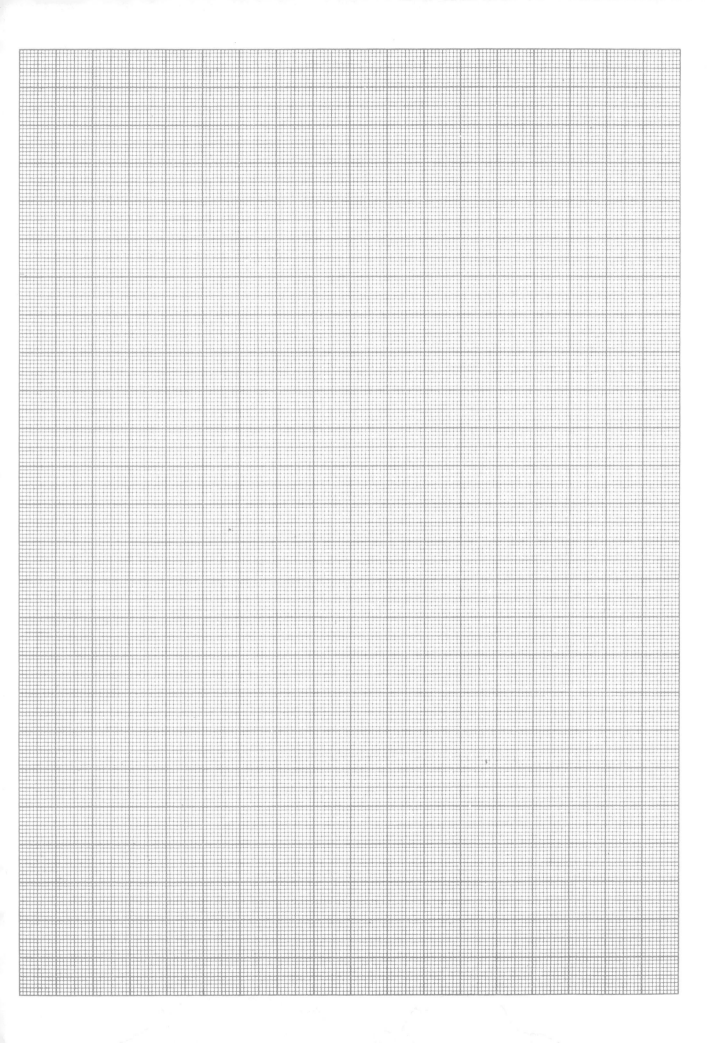

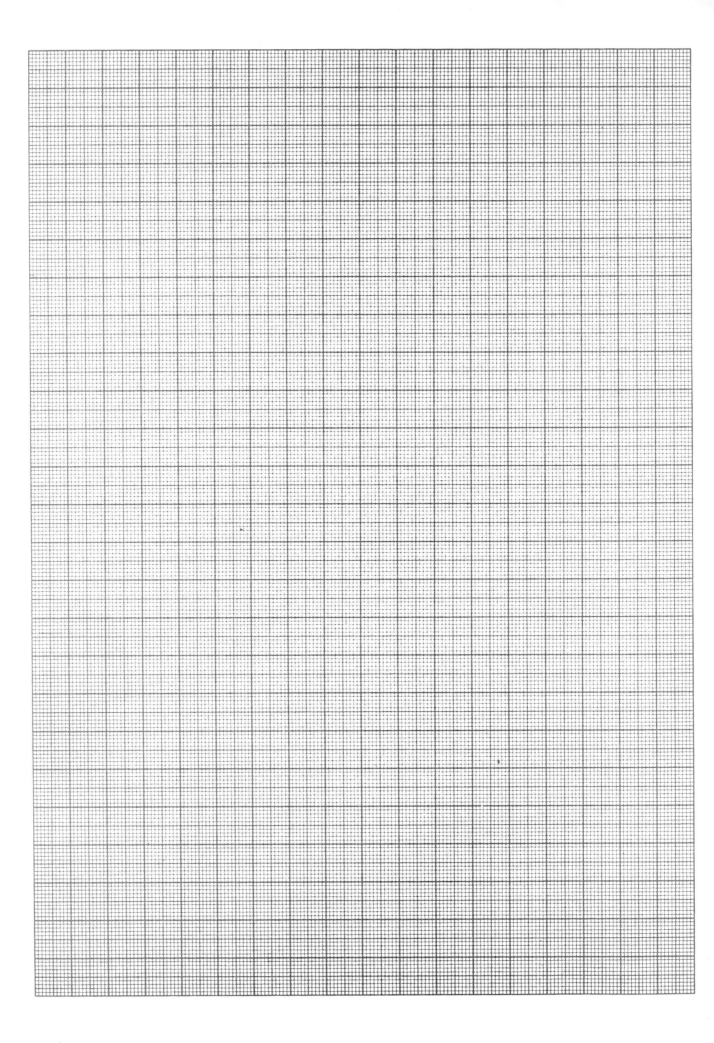

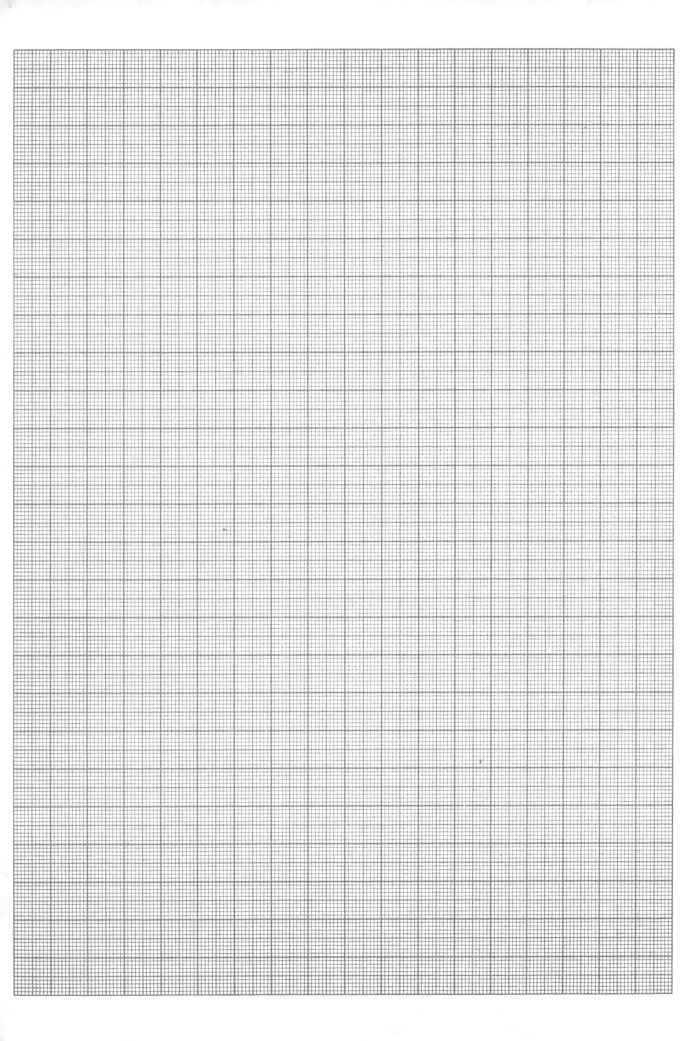

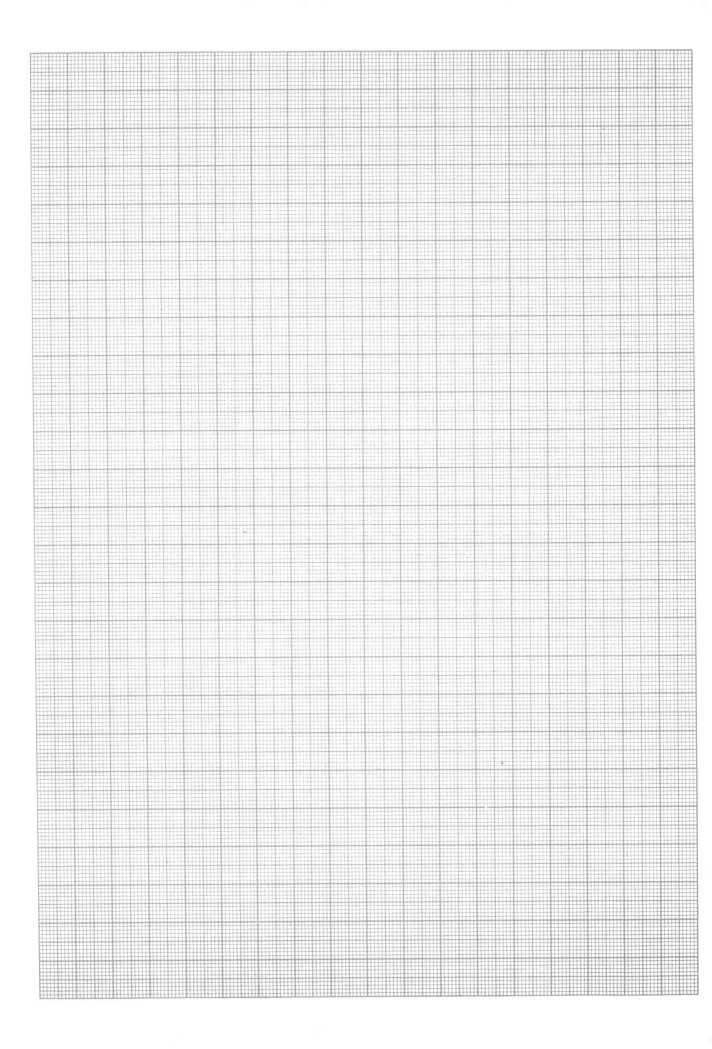

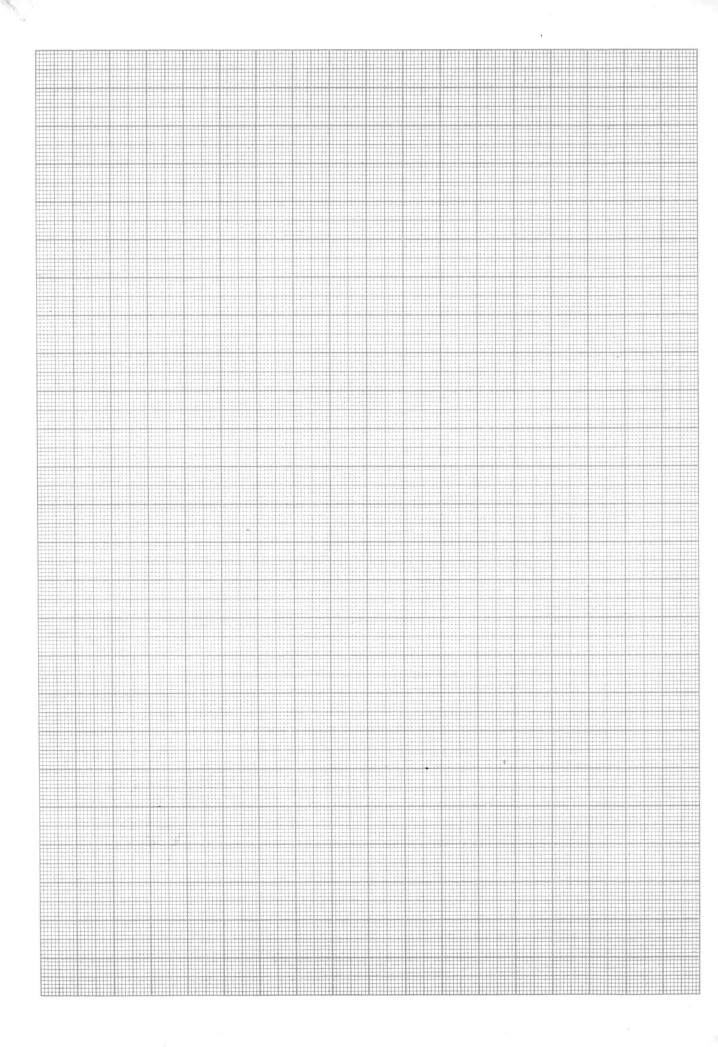